Young Minds Wasted:

Reducing poverty by enhancing intelligence, in known ways.

Third Edition

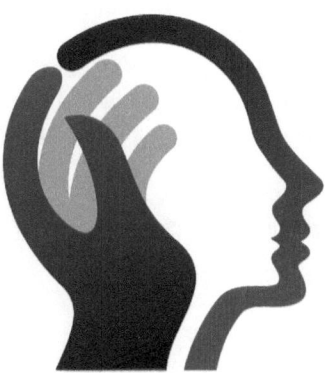

—————— Thomas Schick——

Understanding the relationship between poverty and
intelligence can finally end this age-old circular pattern

Reducing environmental constraints on intelligence, with known interventions, substantially reduces both poverty, and economic inequalities; it
benefits both the advantaged and disadvantaged – society as a whole – and
costs are entirely offset by savings in health costs.

The poor's potential intelligence and capability is painfully and needlessly
wasted. As nutrition and health of the disadvantaged are improved, people
in the developed and under-developed world will more closely reach their
potential intelligence and capabilities – become more productive members
of society.

To Anne, my best friend, life partner, and wife

Contents

Acknowledgements

I am indebted to the patience and editing skill provided by Patrick Inman who helped clarify my thoughts in an earlier version of some parts of this book. I am grateful for the significant advice and encouragement from my friends Ron Levine, a recent North Carolina Director of Health; and Liz Kanof Levine, a previous president of the North Carolina Medical Association; Julian Kitay, a professor of internal medicine and physiology at the University of Texas Medical Branch, whose wisdom and humor led me to incorporate human touch in the book; Ovi de Rivero, previous Peruvian Ambassador to the United Nations, whose global perspective is becoming increasingly relevant; Henry Shaffer, geneticist at the University of North Carolina, whose review significantly clarified the book in many ways including its genetic information; Jeff Soo, among the best croquet players and best coach in the United States – perhaps the world, instilled a more positive view of objectives and potential; Bob Roth, provided a strong sounding board; Rich Burton, Professor Emeritus at Duke University with interests in public policy and business, whose comments significantly improved the book's logical flow; Greg Lockhead, professor emeritus at Duke University, my most warm and respected advisor; my encouraging and favorite brother, Fred Schick, professor emeritus at Rutgers University; my favorite daughter Sarah, and my favorite son Jonathan, both of whom reviewed and helped make the book more readable; and for the probing questions of my very patient life partner, best friend, and wife, Anne.

Prologue

Occasionally, trivial decisions cause dramatic changes in our lives. Some years ago, while my wife and I were working at IBM, I read a notice inviting volunteers to tutor and mentor learning-disabled children. Though I quickly dismissed the request, my life-partner later that evening asked what I thought of the idea. I said that I had no time. "Think about it," she suggested, "It seems exactly right for you." I did, and the next morning signed up. Since then I have worked with over fifty children, always one-on-one, most of them one or two hours a week, some of them for as many as ten years. I got to know them, their parents, and families, and learned a great deal from the children — about them, about me, about society, and about what might work to correct severe societal problems with which we are all too familiar. Experiences with the children drove my interest and enthusiasm. The lives of these children hardly related to anything I had ever experienced. In short order, I learned that they were not as learning-disabled as they were disabled by the baggage they carried; they were not only disproportionately economically disadvantaged but also burdened by an array of severe family issues. I repeatedly noticed that they were apparently bright children, though rarely challenged, hardly motivated, and often lacked self-confidence. I listened to them, and they quickly reacted to my focused attention, sincere interest, and genuine care. It became apparent that for me to help these children, I needed a far more extensive and appropriate background. I signed up for another advanced degree and began work toward a PhD in Cognitive Psychology. Extensive academic studies led me to see the path between their needs and what now seem to be apparent solutions.

The spectrum of issues is considerably wider than was immediately obvious to me. These children hardly reach their potential capability or their potential intelligence. To be clear, while no one does, the advantaged come much closer than do the disadvantaged; limited nutrition and health and other environmental constraints prevent the disadvantaged from getting as close to their potential as is normally achieved by advantaged children. Accordingly, the disadvantaged disproportionately do not complete school, as adults they are less employable, and all-too-often the cycle of poverty continues unbroken for generations.

Furthermore, most of the problems of society such as high school dropout rate, under-employability, criminal activity, and poverty are disproportionally associated with people whose IQ is in the lowest twenty percent of the spectrum. The data are clear. Enhancing intelligence, especially and most substantially of the disadvantaged, would likely significantly reduce many problems of society; it would dramatically increase school graduation, thus employability, and, accordingly, reduce poverty.

Though this book was stimulated by what I learned from these children (unrealized potential, pain, hopelessness, and injustices), the book is about the understanding and knowledge I gained from academic studies and research over a period of more than fifteen years, and what those activities revealed as possible and likely solutions in addressing a wide range of societal problems. This book references and reflects hundreds of academic works, significant research data, as well as occasional anecdotal references to these children. It suggests ways to address many severe social problems in this country, as well as around the world, such as economic inequalities, inequalities in opportunity, limited upward mobility, poverty, and, most relevant to this book, intelligence inequality, that have seemed intractable but, as will be seen, may be substantially reduced, starting immediately.

The problems are most easily touched through the perspective of the children and are most readily understood and resolved with an awareness of solutions which have already been demonstrated. The problems are many, profound, complicated, interrelated, and, most interestingly, when viewed together, present solutions that are far-reaching, straight-forward, and apply, generally, to nearly all the problems, namely, enhance the intelligence in known and proven ways, especially and substantially, of those who are economically disadvantaged.

Directly touching the lives of these children provides a focused and sensitive perspective. Broadening the scope and generalizing the kids' stories to encompass the problems of the wider population allows one to feel the great burdens felt by vast numbers of the economically disadvantaged; the *wasted capabilities* of those in poverty or near poverty, about 150 million people living in the United States, a similar number living in the rest of the developed world, and the six billion disadvantaged people the world over. The wasted human potential represents ongoing severe injustice, mostly of the disadvantaged, and resulting in limitations on the advantaged as well. This book not only probes these basic economic and emotional concerns, as well as their profound implications on societal wellness, but it also presents realistic solutions.

This book recognizes the frustrations and disappointments of the many low middle-class people that are environmentally inhibited from experiencing upward mobility for themselves and their children; people who feel excluded from society's economic advances, who feel their voice is not heard, who know, or should know, that technological advances, such as automation and robotics have and continue to eliminate their jobs while creating new and differently challenging jobs. Their dilemma is quite different from that of those in poverty or near-poverty, but the solutions appear to be the same as for those in poverty; new opportunities through enhanced intelligence and thus additional capabilities. No one should be unnecessarily environmentally inhibited from the benefits of upward mobility; no one's potential should be wasted.

Woven into the fabric of the entire book will be the children's anecdotal stories and events that convey the real life, human sensitivities of far reaching social issues. These stories are linked to relevant research, appropriate data, and analysis. This book conveys not merely a factual base upon which to understand and resolve profound national and international human societal problems, but an in-depth feel for the extent of the problems, a sense of the emotional intensities, and the reality of solutions.

CHAPTER 1
An Overview

"The greatest danger for most of us is not that
our aim is too high, and we miss it,
but that it is too low, and we reach it."

Michelangelo

Poverty has plagued societies essentially throughout all recorded history, and accordingly, many believe it to be a permanent feature of the human condition. While that belief may have been reasonable for thousands of years, it is no longer valid. Today we know that poverty can be substantially reduced, in both the developing as well as the developed world. There is no reason that poverty needs to remain a curse on society effecting the lives of billions of people, not only the poor but the rich as well.

Today, nearly half the world's population, about three billion people, live on less than $3.00 a day. Even more remarkably, half the people in the United States, perhaps the richest country in the world, live in poverty or near-poverty[1]. Some feel, "this is not my problem; it is their problem; they should get themselves out of poverty". Surely, this last comment is appropriate for some of the poor, but not for most; the vast majority of those in poverty are working poor. Some people ask, how can I help? This is the central question. We will return to it throughout this book.

It is generally known that inadequate nutrition as well as poor health typifies poverty. Additionally, as shown shortly, constrained intelligence

---

[1] The term, near-poverty, applies to families whose children qualify for free or reduced-price lunch.

also typifies those in poverty *not because of genetic reasons,* as some might still claim, but because of inadequate nutrition, poor health, and other environmentally determined limitations. Are there relationships between these two observations? Does poor nutrition significantly constrain intelligence? Yes! Professionals firmly agree – the evidence is overwhelming. Does constrained intelligence limit capability, therefore limit employability, and thus, cause poverty? The answer to this question is also a resounding, yes.

We assert and show that a *circular* pattern stubbornly persists; one that can be broken, namely,

> **Constrained intelligence causes poverty,**
> **and**
> **poverty constrains intelligence.**

While, certainly, there are many causes of poverty, *the single most important and widespread cause of poverty is the environmental constraints on intelligence.* Substantial evidence supports this claim; inadequate nutrition and other environmental constraints, typical of the poor, limit the normal development of intelligence resulting in the average IQ of the poor being about 10 points lower than the average IQ of the entire population. Thus, about 75 percent of the disadvantaged have IQs that are below average. We will show that these environmental constraints on intelligence, experienced disproportionately by the poor, can be substantially decreased. Accordingly, intelligence can be significantly enhanced, capability improved, employability increased, and thus poverty can be significantly reduced, perhaps even largely eliminated.

Improving nutrition not only enhances intelligence but also enhances health. For example, providing mineral and vitamin supplements to disadvantaged pregnant women, at pennies a day, not only enhances IQ of the offspring by 8 points[2], it also reduces the frequency of low birth weight (LBW) baby births, potentially saving society several billion dollars every year[3]. Furthermore, providing infants with mother's milk rather than formula not only improves the health of both the mother and her baby but also, according to Quinn and Associates, significantly improves the baby's

---

[2] Harrell and Associates – **The full reference for this and all references are shown in the book's Reference Section**

[3] Initial hospital care of LBW babies costs more than $16 billion in the US alone.

IQ by another 7 points[4]. A meta study[5] published by Horta and Associates in 2015, encompasses 17 studies each of which shows that providing mother's milk increases IQ. This notion is now widely acknowledged, and thus where possible, infants should be provided mother's milk rather than formula.

More generally, there is a strong relationship between environmental constraints on intelligence and health, and thus on medical costs. Many interventions that both enhance intelligence and improve health are widely recognized and not in dispute. These are discussed in Chapters 4 and 5. For perspective, just these 15 IQ points (8 for providing nutritional supplements, plus 7 for providing babies with mother's milk) represent more than one-half the difference between being viewed mentally retarded and average.

Additionally, there is a strong relationship between intelligence and school dropout rate, employability, criminality, and between intelligence and many other societal problems. Not surprisingly, as intelligence increases societal problems decrease as do their substantial costs. Reducing the constraints on intelligence, thereby enhancing intelligence, especially of the poor, dramatically reduces poverty, and many other fundamental societal issues. Chapter 6 expands on these relationships.

Furthermore, there is a strong relationship between one's understanding or, more to the point, one's misunderstanding of intelligence, and various forms of biases which, in turn, influence the perpetuation of poverty and other stubbornly pervasive societal difficulties. So, reducing the environmental constraints on intelligence significantly improves intelligence, thereby not only reduces poverty, but also improves health, accordingly reduces health care costs, and reduces many other problems of society as well.

So, a circular pattern exists not only between intelligence and poverty, but also between education and poverty, nutrition and poverty, as well as between health and poverty.

Constrained health, education, nutrition, and/or intelligence cause poverty,

and

poverty constrains health, nutrition, education, and intelligence.

---

[4] Quinn and Associates (2001)
[5] Horta and Associates (2015); This study is not as directly applicable as is the Quinn study. Chapters 4 and 5 expand on this.

Here, however, we focus on the pattern between intelligence and poverty because that pattern largely explains poverty and also because that pattern encompasses these other patterns, as well.

Charles Darwin wrote nearly two hundred years ago, "If the misery of our poor be caused not by the laws of nature, but by our institutions, great is our sin."[6] At the time, this comment, along with the impact of several of Charles Dickens' books, likely caused guilty feelings, and sensitivity for the poor, but hardly enough change. With the information reported in this book, society may now implement known and understood ways to reduce environmental constraints where they exist so that the disadvantaged may more closely reach their potential intelligence and so, more closely, reach their potential capability. The poor not only endure the pain of poverty, but they experience the consequences of significantly limited intelligence, limited not by genetic factors, but limited by environmental constraints such as inadequate nutrition and poor health.

We need to make several early observations. As indicated by Scarr[7], of the University of Virginia, intelligence is a behavior and, as such, must reflect both cognitive genetic influences as well as environmental influences that include information gained through formal and informal learning experiences. Importantly, inadequate nutrition and other constraining environmental influences constrain genetic influences. Additionally, without adequate nutrition learning is severely inhibited, thus significantly constraining intelligence and its growth. See Chapter 2 for a more complete definition of intelligence.

The good news is that there are many ways to reduce environmental constraints on intelligence, thus enhancing intelligence, especially of the disadvantaged in the United States as well as worldwide. As intelligence in the moment is enhanced, formal and informal learning experiences become more effective, schooling becomes more exciting and appears more worthwhile. Consequently, considerably more disadvantaged children are likely to graduate, *at least* from high school, and as adults are more employable.

More explicitly, because of environmental constraints, the poor are typically less than well-nourished during their developmental stages. Notably,

---

[6] Charles Darwin, *Voyage of the Beagle*
[7] In 1981 and 1998

this not only restrains the expression of genetic cognitive functions through-out the prenatal stage, but, in later stages, limits well-being of the neuronal system, significantly restricting not only intelligence in the moment but also limits the effectiveness of learning experiences – growth of intelligence. Constrained intelligence limits capability which limits opportunities, and thus, all-too-often results in poverty. This, then, summarizes the persistent and stubborn circular pattern – constrained intelligence causes poverty, and poverty constrains intelligence.

Fortunately, we know how to reduce major environmental constraints on intelligence, such as limited nutrition, *at costs to society entirely offset by savings to society.* After reading this, I suspect some people will close this book, exclaiming under their breath something like, *bleeding heart liberal.* Anticipating this, and knowing these claims appear hard to believe, we now show that the cost for supplemental nutrition are more than offset by *immediate* savings in associated areas such as health care. The costs are trivial compared to the short-term savings to society, not to mention the long-term benefits. In net, these interventions are a great bargain.

Many examples, conveyed in this book, support this strong assertion. Mentioning just one far-reaching example here, *inadequate nutrition* not only significantly constrains intelligence but also substantially decreases health, disproportionately increasing the prevalence of major diseases dis-proportionately experienced among the poor including diabetes, high blood pressure, and stress resulting in a wide array of medical conditions.

The most well quantified such medical issue is the frequency of LBW baby births, itself costing the US society about $16.5 billion[8] each year. The disadvantaged have twice as many LBW babies as the advantaged; that amounts to about $11 billion cost per year for the disadvantaged and $5.5 billion for the advantaged. While not valid, let's assume, for the moment, that the frequency of births of LBW babies among the advantaged is ex-plained by natural causes, whatever that might mean. What could explain twice as many LBW births among the disadvantaged? It seems rational to assume, as with the advantaged, that half the frequency of LBW babies among the disadvantaged are similarly caused by natural causes, and that

---

[8] In the United States, 330,000 LBW babies are born each year at an average cost of $50,000 for the initial hospital care of each LBW baby (330,000 X $50,000 = $16.5 billion); the actual cost is much more for disadvantaged babies.

the other half are caused by the disadvantages of poverty. We know that as the nutrition of disadvantaged pregnant women is enhanced by, for example, providing mineral and vitamin supplements, at negligible expense, not only is the IQ of their offspring significantly increased but the frequency of LBW baby births is significantly decreased at great savings to society. These savings support not only the trivial costs that enhance the well-being of disadvantaged pregnant women and thus substantially reduce the incidence of LBW baby births, but these savings also free funds to enhance nutrition and well-being of disadvantaged children through high school. This one example explains how the interventions suggested in this book are entirely cost-free, paid for by savings. Notably, the average IQ of LBW babies is 13 points lower than that of Normal Birth Weight (NBW) babies[9]; improving the nutrition of disadvantaged pregnant women not only reduces the frequency of LBW baby births but substantially improves their NBW babies' IQ.

National health care costs are disproportionately affected by those in poverty. This results not only from poor nutrition but also poor health care and delayed health treatment. These observations are far more severe in the underdeveloped countries. However, as poverty decreases so do much of health care costs. Remarkably, *these significant and relatively obvious observations are hardly pursued by society.*

In short, currently, the disadvantaged, on average, are less intelligent than the advantaged[10]. *As the poor become better nourished and as other interventions are implemented, their intelligent behavior significantly improves. Accordingly, the poor become more capable, more opportunities surface, employability increases, upward mobility increases, and poverty decreases.* This is not hypothetical. The evidence is abundant, clear, and convincing. *Furthermore, as the middle-class grows, more goods and services are demanded, and more profits are made by those providing goods and services. All benefit. Additionally, as intelligence is increased, and poverty is decreased, societal costs for social programs such as welfare are substantially reduced. And we hasten to add, these interventions are entirely self-supporting; as seen just above, costs are immediately far more than offset by savings in medical care and other related areas.*

Some might look at these predictions and conclude this to be a quixotic

---

[9] Sandra Scarr, 1981
[10] Herrnstein and Murray

mission, a quixotic adventure, but evidence for these claims have been reported in juried journals, and we will show these apparently optimistic forecasts to be not only realistic but relatively easily accomplished. We hope the reader will read on to see the evidence for these predictions.

Returning to a previous comment, constrained intelligence is *typical of* people living in poverty. The disadvantaged disproportionately have IQs toward the low end of the IQ spectrum. Herrnstein and Murray show[11] that people with IQs below 75 are fifteen times more likely to live in poverty than are people with IQs above 125; people with IQs between 75 and 90 are five times more likely to live in poverty than are people with IQs between 110 and 125. These observations are much more severe when considering not only those people in poverty but also those in near-poverty, together representing more than half the population of the United States.

By way of clarification, IQ tests measure a subset of intelligent behavior. Constraints on intelligent behavior are determined by both previous inadequate nutrition and limited general well-being, and, accordingly, by less than effective formal and informal learning experiences up to the moment of testing. These are environmental factor constraints; they are hardly affected by genetic factors.

Thus, though we know that those living in poverty are less intelligent, essentially because of environmental constraints, we make the observation that inadequate nutrition and poor health, typical of those in poverty, cause what appears to some as an apparently never ending circular pattern. This pattern, plaguing the human condition can finally be broken.

So, the first set of major messages of this book may be summarized by saying that poverty can be greatly reduced with the realization that interventions can break the above mentioned stubborn circular pattern. It is known, 1) that the poor are typically less than well-nourished and have poor health, 2) that, therefore, the poor are typically less intelligent than the advantaged, and 3) that professionals know that inadequate nutrition and poor health constrain intelligence in all developmental stages. Knowing this, *incredibly, the above rather clear and apparent circular pattern has hardly been recognized.*

The second set of major messages of this book, briefly mentioned above, is a closely related subject with far-reaching societal implications.

---

[11] *The Bell Curve*, pp. 130 - 137

Not surprisingly, evidence shows that as environmental constraints on intelligence are decreased (removed) intelligence is enhanced, societal problems such as poverty, under-employability, welfare, criminality, and school dropout rate are substantially reduced. In the United States, the cost for these societal issues total considerably more than a trillion dollars ($1,000,000,000,000) every year. As shown in Chapter 6, with enhanced intelligence, thus increased capability, increased employability, and diminished poverty, the costs of these societal issues are correspondingly and substantially reduced.

These costs awaken deep rooted ideological debate in governing institutions not only in the United States but throughout the world. They represent just a subset of expenses due to the problems of society; the cost to each individual in society, those directly involved, and those indirectly affected, is hardly measurable.

Figure 1, seen immediately below, conveys the relationship between various problems of society and IQ; the figure shows that many societal problems are associated with people with the lowest IQs. Research indicates, and Figure 1 graphically reports, that most people that are directly involved with various problems of society (such as dropping out of school or being chronically unemployed) disproportionately have IQs toward the low end of the IQ spectrum. As might be expected, with an increase in their IQ, or more generally enhanced intelligent behavior, partly because of enhanced nutrition, they, more likely, finish at least high school, are more employable, upward mobile, and are less frequently directly associated with the problems of society.

Specifically, Figure 1 shows that the frequency of seven different societal problems are progressively lower in each decile of the IQ spectrum; that various problems of society including school dropout rate, employability, criminality, LBW baby births, welfare, and poverty occur most frequently among those having IQs in the two lowest deciles, the lowest quintile (that is, an IQ score below 88) and shows that a relatively trivial percentage of people, having an IQ above 100, are directly associated with these problems of society. While this should come as no surprise, it suggests raising IQ, more generally raising intelligence, especially among the disadvantaged, to reduce the burden of these societal problems.

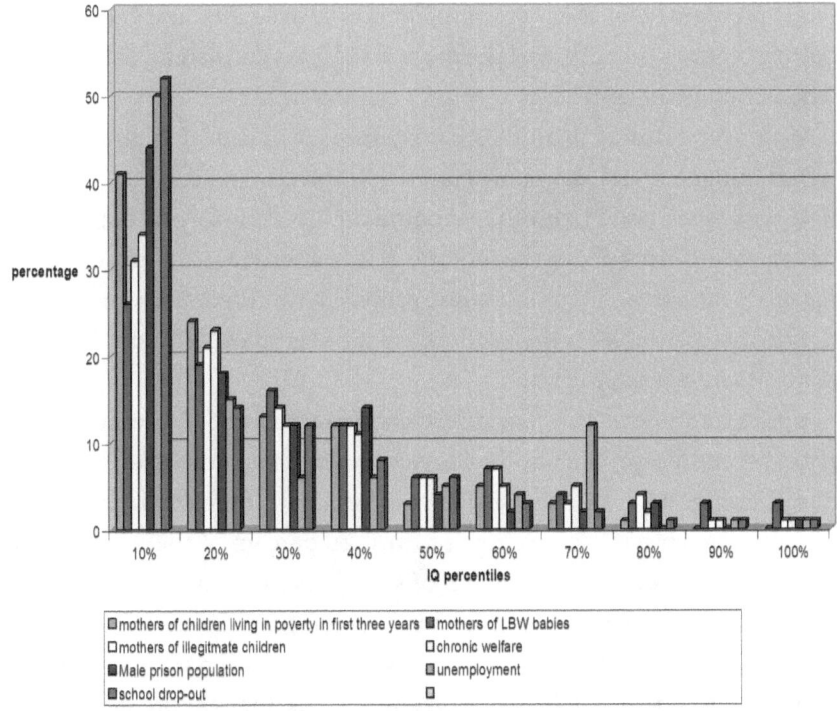

Figure 1. Populations identified with social problems, broken out by IQ decile, based on data from the National Longitudinal Survey of Youth for the age cohort twenty-six to thirty-three years old as of 1990. Adapted from "Social Behavior and the Prevalence of Low Cognitive Ability" in The Bell Curve: Intelligence and Class Structure in American Life *by R. J. Herrnstein and Charles Murray, 1994, New York: Free Press, pp. 383, 378, 381, 377, 372, 374, 376.*

Explicitly, a disproportional percentage, 66 percent, of school dropouts have IQs in the lowest quintile (the lowest 20 percent of the IQ spectrum). 64 percent of chronically under-employable people, similarly, have IQs in the lowest quintile. 62 percent of criminality, also, is associated with those whose IQs are in the lowest quintile (significantly, 74% of prison inmates are school dropouts). So, by improving nutrition and reducing other environmental constraints, especially of the disadvantaged, intelligence is enhanced, health care costs are appreciably reduced, criminality is substantially reduced, and poverty, along with other severe problems of society, are also considerably reduced, at great savings to society. The clear evidence for these assertions is discussed in Chapter 6.

The data shows an associative relationship between IQ and each of these societal problems, but the more associations between IQ and each of these many problems of society, and the more the data is consistent, the more the relationship becomes causal.

Moreover, rational analysis strongly suggests a causal relationship between intelligence, at least as measured by IQ tests, and societal problems. A well-nourished child likely focuses better in not only formal but informal learning experiences. He or she is better able to keep-up in school; education is far more effective. He or she finds school more interesting and is more likely to complete at least high school, be more employable, and thus more likely to move out of poverty.

Figure 1 combines the charts described in Chapter 6. It presents a summary of several key problems of society; it shows that many societal problems indicate similar data; that the bulk of the problems of society are associated with people having IQ scores in the lowest 20 percent of the IQ spectrum. Though the curve for each problem of society is difficult to distinguish in this graphic, the set of curves collectively convey an obvious similarity, an important commonality; the chart shows that these seven societal problems present essentially the same curve. Of course, this does not prove causation, but it contributes to arriving at that conclusion. This can be taken a step further. It is reasonable to conclude that a school dropout is less employable; that's causal. We know that someone who is less employable is more likely to live in poverty and receive welfare; we know these associations are causal. So, while we know the data is associative, providing appropriate rationale, the data is shown to also be causal.

The reason this topic takes up more than an entire chapter in this book is because it presents clear motivation to implement the interventions described. Increasing intelligence, and in turn decreasing many severe and costly societal problems, results in powerful short-term and long-term benefits.

These observations place a much more immediate, much more dramatic, impetus to understand intelligence, and to understand the relationship between intelligence and poverty as well as between intelligence and employability, and various other societal problems; they significantly address many ways to gradually, yet substantially, reduce not only poverty but also other problems of society. Of course, while the resulting potential economic savings are substantial, there are many other cost savings resulting from reduction in poverty felt not only by those directly affected but also those indirectly effected.

Of course, though only some of us are directly involved with these societal problems, every one of us is indirectly affected. These observations regarding strong association between intelligence and poverty, as well as intelligence and other problems of society have profound implications on society and on each one of us.

The broader observation is that environmentally constrained intelligence not only causes poverty but also causes many painful problems of society that not only directly involve and affect the poor but also indirectly effect the advantaged in countless subtle and not so subtle ways. Let's look at this more closely.

Though many of the interventions described in this book show long-term results, they also, and significantly, present compelling immediate benefits or losses. For example, disadvantaged children, likely to be at least somewhat undernourished, are therefore, likely, to demonstrate limited intelligent behavior, and, accordingly, inhibit not only their own learning experiences and that of their disadvantaged peers, but also the education of their advantaged class-mates. Because school systems often teach to the average capability in the class, the advantaged children are prevented from getting as much as they could out of their schooling; they do not get as close to their potential and accordingly their satisfaction and success in life is limited. Advantaged parents are justified to be upset with these educational limitations effecting their children. However, as the disadvantaged become better nourished and healthier they become more intelligent and more capable, essentially immediately. With continuing nutritional interventions their intelligence remains at this more appropriate level, and they remain more capable at every grade level, from early schooling through, at least, high school; teachers can now teach to the higher average, benefiting the advantaged children as well as the disadvantaged. While the benefits described for the disadvantaged appear to take hold in the long-term, they also have important short-term effects. In fact, much of the benefits for both the advantaged and disadvantaged are apparent essentially immediately. Notably, this short-term benefit effecting both the advantaged and disadvantaged is only one of many that are currently hardly recognized by the public, school systems, or vitally, those that determine public policy.

The overarching central message of this book is that while societies have been plagued by poverty apparently forever, that obvious truth need not persist; there are recognized ways to dramatically reduce poverty. Rationally considered questions often lead to reasonable answers. Why do most people in the world who are born into poverty, die in poverty? Does this suggest that one of the major causes of poverty is poverty itself? And if so, what conditions of the poor perpetuate their poverty? In the United States, forty-five percent of those born in poverty die in poverty, incidentally, one of the worst records among developed countries; in England, for example, it's 30 percent, and in Denmark it is less yet.

Though poverty may be significantly reduced in many ways, we propose an approach for the poor to experience upward mobility with no cost to society. We do not suggest doing this by taking from the rich to give to the poor. As Churchill said, "You don't make the poor richer by making the rich poorer." Nor do we suggest accomplishing this by merely reducing the pain of poverty, such as providing welfare. The approach described here is intended to show both the poor and the larger society that the poor may move into the middle-class and beyond by experiencing decreased environmental constraints on intelligence, thereby enhanced intelligence and capability, thus increased employability, and importantly, dignity, self-respect, confidence, and perhaps even happiness.

Though reducing poverty continues to challenge the creativity and ingenuity of peoples around the world, a solution, that knows no national boundaries, has been within reach for a long time. While the next few pages briefly clarify and rationally justify this strong assertion, the purpose of this book is to validate this assertion, to describe its underpinnings, and to surface solutions to severe societal problems.

Although there is no debate that both genetic factors and environmental factors such as nutrition and health determine intelligence, the evidence is unequivocal that limited nutrition and poor health, typical of poverty, significantly constrain the normal development of intelligence. This is widely understood by researchers and academics. Remarkably, though this basic observation is well understood, the resulting step, that constrained intelligence causes poverty, has not been adequately realized or acknowledged by professionals or by the general-public.

Though the public is generally aware that inadequate nutrition and poor health inhibit physical well-being and physical capabilities, for example, in sports, the general-public is less aware that inadequate nutrition and poor health, common among the poor, also constrains intelligence in the moment and thus growth of intelligence as well; inadequate nutrition and poor health constrains the effectiveness of one's neuronal system. Reducing constraints on the development of intelligence, in many known ways, allows people to more closely reach their potential capabilities. This is a fundamental characteristic of the human condition regardless of race or ethnicity. Known interventions substantially improve intelligence and thus substantially reduce poverty. Conveying this truth is a primary purpose of this book.

Some years ago, my wife and I were in Tanzania. We visited several rural schools and learned from the director of one such school that a few years earlier not even 10 percent of the children moved on to the next school level (we would call high school) but since having introduced daily free breakfasts and lunches, a form of gruel, more than 90 percent of the children are moving on to the next school level. Surely, many parents send their children to school for this free nutrition, but the larger result is enhanced intelligence, more effective schooling, and greater opportunities. This has similar benefits, actually significantly more benefits, as free or reduced-price lunch in the United States because these nutritional supplements were for two meals.

To what extent would enhanced well-being increase momentary intelligence, thus, also growth of intelligence, and accordingly capability? Widely acknowledged evidence reveals clear answers to these questions and reveals a path to substantially reducing poverty. Incidentally, we are not saying we can enhance intelligence directly, but rather that we can reduce environmental constraints on well-being, constraints which indirectly restrain and inhibit the normal development of intelligence.

While poverty has many causes including war, drought, restricted opportunities, and low expectations, constrained intelligence, typical of the disadvantaged, is the most fundamental, the most widespread, and the most important. Reducing constraints on the development of normal intelligence, thus increasing intelligence among the poor, would significantly reduce poverty. *And we know how to do that*, and again, importantly, *we know how to do that with costs being entirely offset by immediate savings.* Furthermore, we know that raising intelligence among those with the lowest IQs reduces many severe and costly problems of society. It is constrained intelligence that causes poverty, and poverty that constrains intelligence.

As naive as this may seem to the skeptics, the above is based on clear and undisputed evidence. I hope the doubters will, at least, read through this first chapter, consider the profound and wide-reaching implications, and, at that point feel it worthwhile to consider the more detailed analysis that follows.

Towards substantially reducing poverty in the United States and worldwide

Poverty is defined differently in every country. In the United States poverty is defined by the federal government which recently indicated that families of four, whose income is less than about $28,000, live in poverty; this definition feels different depending upon where in the country one lives. In the United States, the definition of near-poverty includes people whose children qualify for free or reduced-price lunch. This definition, as will be seen in Chapter 5, includes all families with twice the income of those in poverty[12]. Accordingly, about half the US population lives in poverty or near-poverty, that is, are economically disadvantaged.

As intelligence of the disadvantaged increases, poverty substantially decreases. As the poor move into the middle-class, opportunities improve for themselves and their children. Being in the middle-class, they seek more goods and services, benefitting, of course, those already providing goods and services, the already advantaged. Why have the advantaged hardly noticed this progression? Why do many of the advantaged appear significantly indifferent to the pain of the poor? Why don't the advantaged notice the substantial profits they are leaving on the table, significant benefits to the poor, to society, and to themselves?

Nobel Prize winner, Elie Wiesel, wrote on an entirely different subject, "The opposite of love is not hate. It is indifference." Many of the advantaged, not only in the United States but worldwide, are aware of the plight of the poor; much of their taxes support services for the poor, and they give charity that help the poor. Yet, many of the advantaged are largely indifferent to the pain of poverty; the most severe form of indifference being plain lack of interest. Many feel, "The poor take advantage of the system. Why should I

---

[12] Cauthen and Fass (2008), the National Center for Child Poverty

support that?" Some advantaged say that individuals should be responsible for their own lot in life; "Joe got out of poverty. Why don't they? Anyway, why is that my problem?" This indifference, or limited concern, often also results in casually exploiting the poor. Though indifference does not cause poverty, it allows poverty to persist.

Indifference to poverty has become part of the belief system of the advantaged which, historically, has been difficult to modify. However, this belief system has weaknesses – cracks in its armor. By reducing environmental constraints on intelligence of the disadvantaged, thereby improving their intelligence, the poor gradually move out of poverty and into the middle-class; they seek more goods and services. Surely, in short order, as those in poverty move into the middle-class, and beyond, great benefits become apparent to the advantaged; there is reason to predict that the advantaged will become *de*creasingly indifferent. They will sit up and take notice.

A growing schism between the haves and have-nots has contributed and may again contribute to instability, and perhaps, social unrest. This increasing disparity between the advantaged and disadvantaged is not in the best interests of any sector of society. Gated communities—gating some in and some out—likely present conflict and discontent between, and within, both sides of the gates. On the other hand, reducing economic inequalities, by, for example, reducing the environmental constraints on intelligence, thus enhancing capability, and increasing employability, permits more people to move into the middle-class, to participate with and enjoy the fruits of society, to seek more goods and services, resulting in even more entering the middle-class, and the upward spiral gains momentum.

Before turning to another topic, I must respond to an important and anticipated question, that being, how will you prime the pump, how will you pay for the first step of reducing the environmental constraints on intelligence? It's easy to say that once started, the upward spiral gains momentum, but with what impetus will one start the upward spiral? The answer was partly but not entirely conveyed in an earlier discussion. Specifically, providing mineral and vitamin supplements to disadvantaged pregnant women, will not only increase their offspring's IQ, but also substantially reduce the frequency of LBW baby births, saving society many times the trivial cost. So, how will one get the trivial funds to get the ball rolling? I claim society will be overjoyed to make the relatively small initial investment to get the huge and essentially immediate benefits which will lead to the far greater long-term benefits. (In Chapter 4, a simple study is defined towards the end of the section entitled, **The costs of intervention are *entirely offset* by *immediate* savings in related areas,** describing a potential experiment that demonstrates this.)

Interventions that substantially reduce poverty in the United States and worldwide

Several interventions are in place that have already resulted in improvement in intelligence of the poor, at least as measured by IQ tests. Other interventions that enhance capability and well-being, especially of the poor, have been proposed but hardly introduced yet. While the following important interventions are mentioned here, their importance is overpowered by the nutritional interventions already mentioned and to be reemphasized shortly.

1. *early* public-school education starting at age *three,* or even earlier. This would permit disadvantaged children to be better equipped for subsequent formal and informal learning experiences. Additionally, and significantly, this intervention would reduce the number of children identified as learning-disabled[13]. School for three-year-old and four-year-old children should be supported as part of the public-school system with teachers trained accordingly. Laws should be changed to require children to attend school starting at age three.
2. *prenatal and postnatal medical care*
3. parenting training which has a powerful effect on children's intelligence, not only that of the disadvantaged but the advantaged as well. Currently, parenting training is still largely dependent on what was learned from our parents, and we know that that, in many cases, is less than satisfactory. Two examples might make this clear;

> One child, who I had tutored and mentored for several years, at age fifteen decided to go for a joy ride in her mother's car. One of her older friends drove. The girl's mother, knowing her daughter was gone and suspecting what had

---

[13] Muschkin and Associates (2015)

happened, called the police. The girl's mistake, surely a bad choice, had a disproportional effect on her life. Condoning bad choices is not the right response, but there are reactions that work without long-lasting pain. Let the punishment fit the crime, paraphrasing Gilbert and Sullivan in the *Mikado*.

In contrast, another child I knew similarly went for a joy ride with a few friends. He had apparently left his home through a window adjacent to his bed. Upon discovering this at about two in the morning, his mother lay down in his bed and awaited his return. When he stepped back through the window, she grabbed his leg with both hands. He never forgot that moment, that incident, and there was no reoccurrence.

People make bad decisions; children are more likely to do so. Parents, teachers, and psychologists have understood this for many years – parents probably forever. Children are apparently more spontaneous and more impulsive than are adults; those parts of the brain responsible for control remain underdeveloped until the later teenage years. Although we could attend to this known developmental issue, this discussion is left for another time and place. Suffice it to say that different approaches are available to deal with individuals who tend to make poor decisions, particularly those with yet-to-be-developed cognitive control.

But the powerful relationship between intelligence and nutrition, previously mentioned, is far more significant than the combination of all the other interventions just mentioned. Chapter 4 and 5 provide the evidence. The following is a crisp summary of those including, providing,

a. *year-round* free or price-reduced breakfast and lunch not only during the regular school months but during school breaks.
b. free or reduced-price lunches and breakfasts with early schooling.
c. *vitamin and mineral supplements* to disadvantaged pregnant women.
d. encourage providing mother's milk to babies.
e. mineral and vitamin supplements to all disadvantaged school children through high school. Even though the poor in the United States get food stamps and their children receive free or reduced-price

lunches and perhaps breakfasts, evidence clearly indicates that the disadvantaged hardly have adequate nutrition. As will be shown, this intervention alone would prevent the steady decrease in IQ of 0.6 points per year of disadvantaged children.

The costs of these just mentioned interventions are entirely economically covered by savings in the reduced frequency of LBW babies. Some might focus on one or another of these interventions however this book focuses on their aggregate. These interventions result in remarkable and far reaching effects on society. Significantly, again, the cost of these interventions is entirely offset by savings in related areas.

The primary focus, thus, is to show that *an increase of intelligence among the poor results in a dramatic reduction in poverty worldwide.* As this notion is established, understood, and accepted, a foundation is shaped upon which profound societal consequences are likely to be realized.

Increased motivation, enhanced self-confidence, and high expectations raise intelligence

While strong evidence indicates that the major cause of poverty is the environmental constraints poverty places on the normal development of intelligence, other issues influence intelligence as well including motivation, self-confidence, and expectations. Though these capabilities effect intelligence and its growth, reliable quantitative evidence has not been adequately experimentally shown or documented. However, anecdotal stories show that these influencers significantly affect intelligence. The following is one such true story:

> "He cannot do the work", his teacher said to me, *with Charley standing at my side.* When I continued to request the material to help him catch up, she reiterated and said, not subtly, "You do not understand; he *cannot* do the work" clearly implying he cannot and will not be able to do the work. This boy, just entering middle-school, had, years ago, been labeled learning-disabled and thus was placed in a class for remedial math students despite doing well in regular classes for the previous several years. He had been

wrongly placed. It had taken six weeks, with many post-poned meetings, to get the school administration to hold a meeting to reconsider this matter. The administration felt Charley should stay in remedial math class until he proved he could do the work. I argued that he should be placed in regular classes until he proved he could not do the work. Finally, they turned to Charley's mother and asked what she wanted and accordingly it was decided that Charley join regular classes and see what happens. However, he was now six weeks behind. So, Charley and I went to his teacher to find out what topics he had missed. The teacher's message to me, as just mentioned, was not lost on either me or, much more importantly, on Charley, so after we got the material and were walking down the hall, I told Charley that he would prove to her that he *can* do the work, and he did. Within weeks he was up to speed. Incidentally, before the end of the semester this teacher took him under her wing and gave up several lunch hours to help him. This says a lot about her. He had become one of the best, most promising, students in the class.

This true story happens all-too-often; children, especially those that are considered learning-disabled, need an advocate, and they occasionally need help of various kinds. Intelligence is not only enhanced with adequate nutrition, but children need to be given high expectations, motivation, and confidence – they need to know, "I can".

Expectations of the disadvantaged and the learning-disabled are unjustifiably low. Society's expectations of children, in general, our expectations of our own children, are amazingly, and often indefensibly, low. Why do we, parents, teachers, and society in the United States expect so little from our children? Strikingly, expectations of children are much higher in other parts of the world.

These opening pages previewed two sets of major messages of this book; namely, 1) poverty may be greatly diminished by reducing the environmental constraints on the normal development of intelligence, importantly with no cost to society, and, 2) with an increase in intelligence, severe and costly societal problems may be substantially reduced. The following chapters elaborate, clarify, and justify these observations not only backed by juried and often replicated studies but also conveying solutions to a large array of often thought hopeless societal problems. Their costs are far more than offset by essentially immediate health care savings and savings in other related areas. The interventions suggested would substantially reduce poverty and correspondingly increase the middle-class, benefiting all. Additionally, 3) several true anecdotal stories are told which convey the value of motivation, self-confidence, and expectations to increasing intelligence several true anecdotal stories were told which conveyed the value of motivation, self-confidence, and expectations to increasing intelligence. All of this will be expanded in the chapters to follow with widely acknowledged evidence.

One more fundamental message is very briefly introduced now.

Myths, magic, and mystery of intelligence need to be appropriately debunked, demystified

Many myths of intelligence have evolved over the years. These need to be surfaced and dispelled. Phrenology, the study of intelligence by examining bumps on the skull is one such myth. Myths of intelligence have severely delayed and burdened progress toward understanding intelligence. Some of these myths have plagued progress for generations and need, finally, to be recognized as folklore. Perhaps the most important of these is the belief, held by many, that intelligence is largely innately determined. While only *appearing* true for the advantaged, evidence shows that it is not true for the disadvantaged, and, as will be shown, it is not true for the advantaged as well. Intelligence is a behavior, and as such, while certainly partly influenced by genetic factors, is also influenced by environmental factors which must include formal and informal learning experiences. With age the information gathered from these learning experiences occupy an increasing amount of real-estate in the brain which contributes to one's intelligent behavior. The claim that genetic factors are more influential than environmental factors in determining intelligence ignores this steady increase in intelligence and stubbornly contributes to the mythology of the field of intelligence. Because of the importance of this topic to advancing understanding of intelligence, these myths and others are previewed towards the end of Chapter 2, towards the beginning of Chapter 4 and covered more fully in Chapter 7.

These and many other myths have delayed progress towards understanding intelligence. They have caused the study of intelligence to remain mysterious, while understanding has significantly advanced in many other fields. We must develop an understanding of intelligence minimizing its myths, magic, and mystery. Chapter 7 analyzes and debunks a dozen major myths.

Go for the big ones

Psychologists have devoted a great deal of effort improving different aspects of intelligent behavior. Dweck, for example, has shown and documented the importance of the *perception* of one's own intelligence. In contrast, this book focuses on improving the underlying intelligence, the basic and elemental underpinnings of intelligence, by improving nutrition that enhances the effectiveness of the neuronal system, cognitive structures, and formal and informal learning experiences. Dweck's findings are more effective with observations demonstrated in this book.

As a variation on the words of Michelangelo stated earlier, former Israeli President Shimon Perez said that ***effecting small changes takes about the same effort as effecting great changes; go for the big ones***. Here, we focus on a big one, on several big ones, and not so amazingly, with one solution. Often one tries to solve only one sliver of a problem at a time because that seems to permit focus and a bounded set of considerations. But some problems are perhaps better understood and analyzed by examining a broader perspective, thus potentially recognizing interrelationships that would not have otherwise become evident. Analyzing the wider scope of problems often reveals patterns and often suggests the questioning of notions that may have become too comfortable in more narrow analyses.

In this book, upon probing the commonalities of a broad set of societal concerns, we seek solutions to the problems of upward mobility, economic inequalities, inequalities of capabilities, inequalities of opportunity, the stubborn intelligence gap between the disadvantaged and advantaged, ways to reduce poverty, approaches that reduce the cost of medical care, and ways to reduce welfare, criminality, and other problems of society—a tall order indeed. We assert, however, that these problems, each with enormous complexity, and appearing disparate, have a common underlying resolution, one that has recently become better understood. The challenge, here, is to step back from the disjointed narrative, to take-in and experience the bigger picture, to seek understanding of the whole as well as its pieces, and to recognize a basic human strength that can be far better realized to correct profound and demanding problems. *That basic human strength, essentially always under-realized, is human intelligence; more to the point, this basic human strength, among the poor, is all-too-often largely under-realized.*

After reading this chapter some may claim these notions to be simply theoretical, but the evidence is not in dispute. Unequivocal data shows that the economically disadvantaged are less intelligent than the advantaged, that environmental constraints limit the intelligence of the disadvantaged, that there are demonstrated ways to decrease these constraints, thus, known ways to improve intelligence, especially and most substantially of the disadvantaged, and accordingly, as intelligence of the disadvantaged increases, so does capability, opportunity, employability, such that the path of upward mobility opens-up, and poverty decreases. This is not theoretical but rational, reasonable, and waiting for acceptance. Additionally, as the IQ, more generally the intelligence, of the disadvantaged increases, costly and stubborn societal problems substantially decrease including school dropout rate, under-employability, criminality, and welfare services.

The next chapter, Chapter 2, presents a platform upon which to build an understanding of intelligence. Chapter 3 examines the current understanding and misunderstanding of poverty. Chapters 4 and 5, describe ways to reduce the environmental constraints on intelligence and quantify the benefits that essentially eliminate the IQ gap between the advantaged and disadvantaged, regardless of race. Chapter 6 conveys that as intelligence is increased societal problems such as school dropout rate and welfare are reduced. Chapter 7 shows that the myths, magic, and mystery of intelligence are still alive and well; the myths of intelligence must be surfaced, dispelled, and finally safely placed into the mythology of the field. The final two chapters summarize the book's messages and reviews directions to largely eliminate poverty.

Would increasing intelligence and capability support the mobility of disadvantaged people to move into the middle-class? Evidence shows the answer is an unqualified, *yes*. Societies have been plagued by poverty likely from prehistoric times, with many thinking that poverty is a permanent feature of the human condition. That need not be true. Breaking the historically stubborn circular pattern – poverty causes poverty – by reducing environmental constraints on intelligence and by increasing expectations, motivation, and confidence, is what this book is about. Poverty can be substantially reduced, largely eradicated, in the United States, and worldwide.

A friend recently asked, "Why have I never heard this rather obvious connection between constrained intelligence and poverty" I told him that the array of societal issues discussed in this book are usually viewed separately because they are, *apparently*, so diverse and complex. In contrast, I said, we looked at these apparently disparate problems with an encompassing view, discovering that they are really closely related; that removing environmental constraints on intelligence, disproportionately experienced by the disadvantaged, dramatically increases the capability of the poor, increases their likelihood of completing school, increases their employability and in better jobs, substantially reduces opportunity inequalities, significantly reduces economic inequalities, considerably reduces poverty, welfare services, criminality, essentially eliminates the stubborn intelligence gap between the advantaged and disadvantaged, and also markedly reduces cost of medical care. Yes, these deceptively different, dissimilar, and stubborn societal problems are intimately related and are best understood and resolved when analyzed together. We have done just that.

We now proceed to demonstrate how these assertions are justified.

CHAPTER 2
A View of Intelligence

Mark Twain is believed to have said,
> "When I was a boy of fourteen, my father was so igno-
> rant I could hardly stand to have the old man around.
> But when I got to be twenty-one, I was astonished at
> how much he had learned in seven years."

Al Peeden, my friend and barber said it perhaps more mem-
orably. He said,
> "It's amazing how much I learned since I thought I
> knew it all."

Several threads, already introduced in the tapestry of this book, are rein-
forced to strengthen, provide cohesion, and assure durability of the entire
tapestry. Toward that end, this chapter includes research findings that provide a foundation to understand wide-ranging topics of intelligence; these
studies provide credibility of the conclusions developed. I have woven several anecdotal stories into various parts of this book to convey the warmth,
appropriate sensitivity, and humanity of the issues. This information is not
only revealing but often inspiring and surprising.

   We talk of human intelligence as if we understand what it is, how it is
determined, and how it may be measured. Though a great deal of research
has been devoted to understanding intelligence and intelligent behavior,
and while significant progress has been made, an understanding of intelligence or, more specifically, an understanding of human intelligent behavior
remains elusive. Why do we claim to measure intelligent behavior with IQ

tests, when we know these tests do not measure important aspects of intelligence such as creativity, people sensitivity, and artistic capability – when we know that no scientific approach has ever been defined to measure intelligence – when we know that we have no way to measure the collection of characteristics we call intelligent behavior? Why do many, all-too-often, assume IQ tests measure innate intelligence when we know that currently we have no way to measure that? Why can we place people on the moon and bring them back safely, yet, arguably, the most central feature of the human condition, intelligence, eludes comprehension. Answers are not obvious; perhaps the difficulty stems from being too close to the problem; perhaps we all seek understanding through differently disabling filters, or perhaps we have been seeking answers in all the wrong places (paraphrasing the comedian Eddie Murphy, on an entirely different subject).

Terminology

Researchers employ differing concepts of intelligence.[14] In the literature, the term *intelligence* is sometimes used interchangeably with the phrases *intelligent behavior*[15], *innate intelligence*[16], and *IQ*. Inconsistent, contradictory terminology often leads, even professionals, to talk past each other with some believing intelligence is stable and others defining it as dynamically changing and growing with experiences. To avoid confusing measured and inferred phenomena, this book often employs the following narrowly qualified phrases in place of the word *intelligence*. Notably, while new definitions are avoided, they are introduced, as required, to refer to new or more encompassing notions.

- *Intelligent behavior* demonstrates "the ability to understand complex ideas, to adapt effectively to the environment, to learn from experience, to engage in various forms of reasoning, and to overcome obstacles by taking thought"[17]. This is a widely used definition even

---

[14] Fundamental scientific concepts often defy definition. For example, while we know how to create, use, and measure electricity, the word *electricity* remains undefined.

[15] Scarr, 1998

[16] For example, Burt, Jones, Miller, and Moodie, in their book published in 1934

[17] Neisser et al., (1996), p. 77; cf. Gottfredson, 1997

though difficult to scientifically explain or measure. Furthermore, this definition does not reflect musical talents, artistic capabilities, sensitivity to or for others (people sensitivity), creativity, or perceptivity.

Additionally, all the phrases in the above definition seem to deliberately omit any mention of knowledge. This is most obviously exemplified in the phrase *to learn from experience* implying that intelligent behavior only includes the learning process but not that which is learned – knowledge.

While some psychology professionals claim intelligence does not include knowledge, this claim is absurd because intelligence appears, is manifested, as a behavior, and as such, must reflect all factors that influence intelligent behavior including knowledge. Surely, if intelligence were characterized by only genetically determined structures, essentially established at conception such as cognitive structures, then, clearly, intelligence has never been measured. More to the point, knowledge increasingly occupies the brain as experiences are accumulated; they increasingly determine intelligence and intelligent behavior across most, if not all, an individual's life; knowledge determines the growth of intelligence. Knowledge has traditionally been omitted from the definition of intelligence apparently because hereditarians have influenced the definition; from the hereditarians view only that which is genetically determined may be defined as intelligence. We reject that notion.

Also, as mentioned earlier in the book, intelligence is not stable or fixed as some have claimed. One's intelligent behavior changes from one instant to the next according to environmental factors encountered, such as nutrition, health, and learning experiences. Adequate nutrition and health allow one's brain, including cognitive functions, to perform closer to peak potential in the moment. Intelligence depends on the performance of one's neuronal system in the moment, but it is also determined by the effectiveness of processing formal and informal learning experiences and thus, contrary to widespread opinion, intelligence is essentially open-ended—the more effective the learning experiences, the more knowledge, the more intelligence.

Thus, we use a more encompassing definition of intelligent behavior, namely, "the ability to understand complex ideas, to adapt effectively to the environment, to learn from experience, to engage in various forms of reasoning, and to overcome obstacles by taking thought", all reflecting accumulated knowledge, and reflecting musical talents, artistic capabilities, sensitivity to and for others (people sensitivity), creativity, and perceptivity.

- *Intelligence* is a characteristic of the mind whereas *intelligent behavior* is its demonstration – its manifestation. It results from both genetic and environmental influencers. The words *intelligence* – a trait of the mind – and *intelligent behavior* – its manifestation – are recognized as referring to different notions. However, as intelligence of the mind cannot be measured except through its manifestation, and because of the historic and continuing misuse of the words, they are occasionally used interchangeably. Care is taken so that when the word intelligence is used to refer to a characteristic of the mind, that usage is highlighted.

- *The brain, consisting of billions of neurons, contains genetically determined structures including cognitive facilities, memory facilities, interpretive facilities such as those associated with hearing and sight, and a universal grammar. The workings of the neuronal system(s) appear better understood than the workings of the mind. The genetically determined memory system(s) start existence as a blank slate; formal and informal learning experiences – environmental influences – determine the memory system(s) contents.*

- *The mind* uses the neuronal system(s), structures, of the brain to accumulate, analyze, store, organize, process, as well as access information. Creativity, decision-making, as well as other basic functions of the mind make use of accumulated information. Hypotheses have been developed intended to describe the workings of the mind which seem to rely on our current understanding of how computers operate. This may prove productive but, in my opinion, when we learn how the mind really works we will have learned a greatly improved way for computers to work.

- *Inhibited or constrained intelligence* is intelligence that has only been partially realized because of limited nutrition, health, and, accordingly, causing less than effective formal and informal learning experiences. These limited environmental factors disproportionately affect intelligence of the disadvantaged in every stage of life.

- *Innate intelligence* refers to that part of potential intelligence determined at conception. Currently, it is not measurable, and while theories of how the brain works have and are being developed, we know relatively little about innate intelligence. The term is rarely used in this book.

- *Momentary intelligence* refers to intelligent behavior observed in the moment. Momentary intelligence constantly changes according to well-being, experiences, and emotional state.

- *Potential intelligent behavior* refers to the level of intelligent behavior an individual might achieve given an innate intelligence, adequate well-being, and optimal growth of intelligent behavior. Potential intelligence is influenced by genetic factors, which environmental influences may constrain (for example, by inadequate nutrition) as well as enhanced (for example, by learning experiences), and one's chosen activity. For example, if the chosen activity is right for an individual, such as music for, say, Mozart, he will likely behave more intelligently and creatively than if he, Mozart, chose politics.

  None of us achieve our potential intelligence; some of us get closer than others. Those that are and have been consistently healthy and well nourished, and thus have had a steady state of well-being, and have had correspondingly effective formal and informal learning experiences, come closer to their potential intelligence than those who have not experienced well-being. It is safe to say that every person has experienced environmental limitations that have inhibited him or her from achieving his or her potential, and this is far more true for those who have experienced many and severe environmental limitations such as the disadvantaged. Reducing environmental constraints would reduce the inequalities of opportunity, and likely narrow the intelligence gap between the haves and the have-nots.

- *IQ* refers to a measure taken of a subset of capabilities of momentary intelligent behavior.

Among the aspects that intelligence reflects but IQ hardly reflects are creativity, artistic capability, musical potential, and people sensitivity[18]. IQ must not be equated to intelligent behavior, though it all-too-often is.

We all know intelligence when we see it, and much of that realization hardly relates to characteristics measured by IQ tests. Significantly, though IQ tests measure a meaningful slice of intelligent behavior, they hardly measure the full scope of intelligent behavior. While this book frequently refers to this more encompassing view of intelligent behavior, it often focuses on the subset of intelligent behavior measured by IQ tests (most often Wechsler Intelligence tests), because these tests are the operational measures used in many studies.

Intelligent behavior may be enhanced or diminished according to nutrition, health, formal and informal learning experiences, and other environmental factors; it varies from moment to moment. Currently, only momentary intelligent behavior is measurable. The phrases *momentary intelligence, intelligent behavior, and intelligence may on occasion be used interchangeably, but in this book, they* are not used interchangeably with the phrase *innate intelligence, potential intelligent behavior,* or *IQ.*

While innate intelligence is determined at conception and, for the most part, typically, does not change during the life of an individual, momentary intelligence, intelligence, or intelligent behavior changes with every environmental experience encountered throughout life. Interestingly, though we currently lack a means to measure or enhance innate intelligence, we have many ways to both measure and enhance intelligent behavior.

It is important to distinguish between maintenance of intelligent behavior in the moment—reflecting the factors that enhance or detract from an individual's performance at a given time, and growth of intelligent behavior. Both are limited by the individual's well-being. For example, inadequate nutrition limits a child's (a person's) capability in the moment. It also causes a child to pay less attention in class or informal learning experiences, thus acquiring less knowledge and accordingly performing less well on subsequent

---

[18] Howard Gardner, (1987)

activities. Thus, poor nutrition affects both momentary intelligence and growth of intelligence.

While IQ scores are age-independent, intelligent behavior is age-dependent.

IQ measures are designed to be stable throughout the developmental stages; they are designed to be age-independent. In contrast, intelligent behavior surely increases with age; six-year-olds behave less intelligently than do sixteen-year-olds. Stated differently, while the IQ of an individual is designed to remain constant throughout most of his or her life, intelligent behavior changes with every experience, changes frequently according to well-being, and, importantly, changes according to formal and informal learning experience. IQ scores may only remain stable—age-independently, if intelligence grows at every age—age-dependently, age-appropriately. For example, an individual's IQ may remain stable if his or her vocabulary and math skills increase age-appropriately.

While the IQ of the advantaged (the data focuses on whites) remains stable over time, the IQ of the disadvantaged (the data focuses on blacks) gradually and steadily drops every year of the developmental stages. The reports of several researchers including Nisbett, Jensen, Dickens and Flynn, have made this observation. Its recognition helps explain why blacks, on average, have lower IQs than do whites, at each age level (see Figure 5 for details). The specifics show that the IQ gap between blacks and whites is about 4½ points at age four increasing, notably, at a steady rate of about 0.6 points every year to age twenty-four. Genetic influencers such as puberty have not surfaced that may explain this. As shown in Chapter 5, the gap is environmentally determined—the gap is entirely caused by constraints such as nutritional well-being so that formal and informal education is less than effective, causing a steady drop in IQ.

In 2015, Stumm and Plomin, disregarding race, documented a study showing that SES significantly affects intelligence; the study showed that this was not a racial issue.

This study showed that children from lower SES backgrounds tend to perform on average worse on intelligence tests than children from more privileged homes as early as at the age of 2 years. Furthermore, SES accentuated these differences throughout childhood and adolescence: the 6-point IQ difference in infancy between children from low and high SES homes almost tripled by the time the children were 16 years old. Our [they wrote] findings confirm changes in intelligence throughout early life and suggest a meaningful relationship between IQ growth and socioeconomic factors.

Here again, the data correspond very closely to the above data for blacks suggesting that the issue is not racial but economic disadvantage.

That part of intelligent behavior measured by IQ tests can be significantly enhanced across the entire spectrum but most dramatically for those scoring in the low end of the spectrum. As strange as it may seem, raising the IQ of those having an IQ below 87 by an average of ten points would raise the IQ of most of the people in the lowest quintile (those with scores below 87) of the IQ spectrum to above 87.

There will always be a lowest quintile of the IQ spectrum, but the IQs of those currently in the lowest quintile could be significantly raised and thus would consist of people far more capable to experience and participate in the fruits of society.

Intelligent behavior is not stable, but grows according to experiences

As indicated just above, while the IQ of whites essentially remain stable over time, the IQ of the disadvantaged gradually and steadily drop every year of the developmental stages. How may this be explained? Though, currently, genetic factors cannot explain this phenomenon, inadequate nutrition and other constrained environmental factors do.

Intelligent behavior reflects not only genetic factors but also accumulated information. In his 1990 work, Ceci writes (p. 117):

> many theorists distinguish between knowledge and intelligence; the latter has historically been conceived of as the underlying capacity to acquire knowledge, detect relationships, and monitor ongoing cognitions in order to adapt to

ever-changing demands. Thus, according to those researchers, intelligence refers to one's aptitude for using cognitive
processes in situations where knowledge either is of minimal importance or is so basic that it is thought to be shared
by all persons (Gregory, 1981). [On the other hand, if] one
wishes to locate the most complex thinkers at a racetrack,
the soundest advice is to begin by testing for factual knowledge about racing. High levels of cognitive complexity at
the races are almost never found among unknowledgeable
patrons, regardless of how high their IQs may be (Ceci and
Liker, 1986a, 1986b; 1988; Liker and Ceci, 1987). This is in
contrast with the view of *intelligence* as "inborn, all-around,
intellectual ability" indicated by Burt and his coauthors in
1934. Burt's view of intelligence excludes acquired information and strategies gained from experience; intelligence,
he relates, is entirely innately determined. Importantly, this
view of intelligence is unfortunately still held by many educators, psychologists, and many in the general public.

The operational measure of intelligent behavior used throughout the
literature, and assumed valid in this work, is IQ testing: testing that measures not only cognitive capabilities but also reflects knowledge gained from
formal and informal learning experiences including, at least, vocabulary,
strategies, and math skills. Intelligent behavior, as measured by IQ tests, in-
corporates age-appropriate knowledge gained over time. Of course, IQ testing has severe weaknesses; it does not reflect creativity, artistic or musical
orientation, or people skills, all of which contribute to intelligent behavior.
But while IQ testing remains among the most widely accepted measures of
intelligent behavior that is used today, there is reason to believe its use is and
will continue to decline.

Intelligence tests currently cannot isolate cognitive genetic factors that
influence intelligent behavior; we do not know how to measure those factors directly. Intelligence tests measure behavior and, as such, reflect not
only genetically influenced intelligent behavior but also reflect intelligent
behavior as environmentally influenced by nutrition and health—well-being—*and also* reflect intelligent behavior as environmentally influenced by

formal and informal learning experiences. The tests are designed to account for the fact that intelligent behavior is age-dependent, which grows partly due to learning experiences. (Intelligence test scores also indirectly reflect self-confidence, emotional well-being, motivation, and expectations— factors that are hardly addressed in this part of the book.)

Several studies show that IQ is stable; the average IQ of a group of six-year-olds correlates very strongly (0.70) with the same group of individuals at age sixteen. Jensen then of Stanford University, reported this in his writings of 1969 and again in 1998. In support, Pinker, a Harvard professor in psychology writes in his 2002 book "there is now ample evidence that intelligence is a stable property of an individual" (p. 150). This statement is an example of the ambiguity that arises from the use of the word *intelligence*. While IQ may be relatively stable for those consistently and adequately nourished, intelligent behavior is not stable but changes with formal and informal learning experience and every environmental change. The disadvantaged are likely to be less than adequately nourished and accordingly probably gain less from formal and informal learning experiences. This explains why their IQ does not remain stable but rather gradually and steadily decreases over time.

Snyderman reported in 1988 (pp. 81–82) that Binet, the founder of IQ tests, "believed that intelligence, as measured by the [IQ] tests, could be improved, and… criticized that 'brutal pessimism' of those who would forever consign the backward child to a subnormal life." Binet repeatedly conveyed that IQ scores are highly malleable. He criticized the misperception that test results determine later intelligent behavior. That criticism remains relevant today.

Intelligent behavior must increase age-appropriately for IQ to appear stable.

Chomsky first reported in 1975 that a *universal grammar* is genetically determined, innately determined, for all humans. This notion is now widely acknowledged. However, IQ tests measure one's vocabulary, which is unique to particular cultures; unlike the universal grammar, vocabulary is known to be environmentally determined by formal and informal learning experiences—surely, specific languages are not genetically determined. IQ tests measure some aspects of one's reasoning ability, including aspects that may be genetically determined as well as environmentally learned. They also measure other learned skills such as strategies and vocabulary, capabilities that are entirely environmentally learned.

The ability to accumulate knowledge from formal and informal learning experiences is limited by poor nutrition; subsequent access to this accumulated information is also affected by limited well-being. Even assuming exposure to *equal* formal and informal learning experiences, individuals who are poorly nourished gain less from experiences than do those who are adequately nourished. Additionally, as Sigman and Whaley explained in 1998 (p. 176), "Children who grow up hungry are likely to have less sense of security and well-being [than those] who do not... This psychological sense of well-being is likely to affect children's ability to reason and learn."

Knowledge gained from learning experiences is cumulative through much of an individual's life. Moreover, new material is more readily understood if one has previously encountered similar or related material. Formal and informal learning experiences significantly affect intelligent behavior and differentially affect different groups—for example, the adequately nourished and the inadequately nourished, the economically advantaged and the economically disadvantaged.

Intelligent behavior (as measured by IQ tests) can be significantly enhanced across the entire IQ spectrum but most dramatically for those scoring in the low end of the spectrum. As mentioned, raising the IQ of those having an IQ between 55 and 87 by an average of only ten points, would raise the IQ of all those scoring in this range, in the lowest quintile of the IQ spectrum, to above 87[19].

---

[19] Notably, according to the standard IQ normal curve, while about 2,000 people out of 10,000 would have an IQ below 87 (about 20 percent of the entire population, the lowest quintile) 228 people out of 10,000 would have an IQ below 70, and about 13 people out of 10,000 would have an IQ of less than 55. We therefore only consider that population with IQs above 55. Moreover, about 12 people out of 10,000 have Down syndrome and thus the IQ bell curve does not follow the standard normal curve exactly, but rather a bubble exists at the lowest end of the IQ spectrum reflecting the incidence of individuals with Down syndrome and other genetic conditions limiting intelligent behavior.

To raise the intelligent behavior of those currently having IQs between 55 and 8 to the level currently associated with an IQ of 87 would require raising their IQs by a range of 1 to 32 points, though it would require raising the average by about 10 points because of the IQ normal distribution just described. And while raising an IQ by 32 points might seem unlikely, it applies to a relatively tiny percentage of the distribution. Only

There will always be a lowest quintile of the IQ spectrum, but those in this raised lowest quintile would consist of people far more capable to produce and participate in the fruits of society.

Several conclusions can now be made that are basic to understanding intelligence. First, while IQ scores may remain stable given age-appropriate learning experiences and given adequate environmental experiences, *intelligent behavior is not stable; it grows with formal and informal learning experiences.* Intelligent behavior is age-dependent (grows in childhood with age) while IQ scores are designed specifically to be age-independent (remain stable as one ages). Second, growth of intelligent behavior, of course only recognizable over time, is separable from momentary intelligent behavior. Third, without growth of intelligent behavior, IQ scores drop over time; *for IQ scores to remain stable, intelligent behavior must grow age-appropriately.* More will be indicated about these observations in Chapter 7 entitled "Myths, magic, and mystery of Intelligence".

A view of intelligence

Human intelligence is based on several genetically determined capabilities such as cognitive functions, a universal grammar, and a memory system including the storage, integration, analysis, and access of information. Several other genetically determined capabilities determine intelligence such as distance and direction perception, as well as visual and audio input interpretation. It is also safe to observe that genetic factors determine, and environmental factors influence, these capabilities. Many species, not only humans, enjoy many of these capabilities.

Because the genetic makeup of every human is unique one might reasonably conclude that the intelligence of every human is also genetically unique. But that need not be so. It could be that intelligence of each human being is

---

0.022 of the entire population has an IQ below 70; 0.18 of the entire population has an IQ between 70 and 87. So, about 1/9 of the people in the lowest quintile have IQs below 70. Enhancement of the intelligent behavior of the vast majority of that entire quintile is probable; the most dramatic enhancement is likely for those with IQs below 70. More on this later.

Note, this begs the question of the periodic renormalization of IQ tests. The distribution of human height and weight are allowed to change freely. Why not the distribution of IQ scores? This is discussed more fully in Chapter 7.

unique not only because of genetic differences but perhaps largely because of environmental differences, most importantly, because of differences in the contents of our memory, formal and informal learning experiences; unique learning experiences make the intelligence of each human unique; our brains are not interchangeable because of these unique experiences, which in fact, determine who we are and how we behave. Motivation and stress are arguably genetically determined though, surely, environmentally influenced.

A philosophical consideration

Many people seem to take great pride that they, as individuals, are unique. In fact, as mentioned earlier, no two individuals in nature are identical; for the most part we are magnificently similar. Our heart, lungs, blood, liver, and much more are sufficiently alike to be essentially interchangeable.

Why, then, do we so strongly emphasis our differences? Why do we stress, and stress over, our differences in race (whatever that means), in culture, capability, and in talent? Our similarities are overwhelming, and most of our differences may be just that, simply differences. So, again, why do we persistently emphasis our differences.

While much of our anatomy is interchangeable, our brains are not interchangeable. But what of our brains makes us unique? What of our brains are genetically similar, and what are genetically different? Currently, we hardly know enough to say, though there is reason to conclude that while the cognitive processes, memory processes, universal grammar, and perception processes, are likely genetically much the same, the information in each human's memory system is unique because it is not genetically determined but rather environmentally determined according to each individual's unique emotional as well as formal and informal learning experiences. This observation is vital to our understanding of human intelligent behavior because it conveys that both emotional and learning experiences—environmental factors—significantly determine differences in our brains, and differences in our intelligence.

A philosophical question remaining unanswered is, while we know that when we transplant a heart, a lung, or some blood from one person to another, the recipient remains the same person, whereas, if we were to be able to transfer a brain from one person to another, we continue to question who the recipient is. Is he or she the donor or the recipient? And that question is interesting because our brain and its memories determine who we are and how we behave. This author agrees with those that have concluded that after a brain transplant is successfully completed, the recipient becomes the donor, albeit with a new envelop.

Interventions that enhance intelligence

Partly because intelligence is less mysterious than it has been, we now know how to indirectly enhance intelligent behavior, as is briefly shown and expanded with a great deal of evidence in Chapter 4. We also know that the IQ gap between the disadvantaged and advantaged, and between blacks and whites, can be essentially eliminated, as shown quantitatively in Chapter 5. Throughout this book several deeply held notions are discredited and shown to be myths. Chapter 7 focuses on myths that have severely delayed progress towards understanding intelligence.

So, a purpose of this book is to answer the question, can reducing environmental constraints, thereby indirectly enhancing intelligence, most substantially and especially of the disadvantaged, improve upward mobility, significantly reduce poverty, decrease the problems of society, reduce economic inequalities, and prevent the development of a two-tier society? And, of course, these issues apply not only to the United States and the developed countries, but also to the under-developed countries of the world.

We start with studies that have shown the effect of environmental influences on intelligence. For example, as Jensen, then of Stanford University, reported, identical twins, on average, have a 6-point lower IQ than single births. Sharing the mother's resources prenatally and the parents' postnatal attention—both environmental considerations—determine these differences. The heavier twin at birth has an average IQ of about 5.5 points higher than the lighter twin. That means that the lighter twin has an IQ 8.75 points lower than an average single-birth child. How can these differences be caused by genetic considerations when the twins resulted from the same fertilized egg? Placing these numbers in perspective, a person labeled mentally retarded has an IQ 30 points below someone who is considered average in intelligence; 8.75 points is a big deal.

Additionally, Frank Sulloway, at the University of California, Berkley, reported in 2007 that first-born children have an IQ 2.3 points higher than the second born, whose IQ is higher than the third born, whose IQ is higher than the fourth born, and so forth. Genetics does not explain these differences. The environment does. Stephen Ceci, of Cornell University, indicated in 1990 (p. 77) that every year of missed high school results in a 1.8-point loss in IQ. Jencks and associates[20] showed in 1972 that educational summer activities prevent the loss of IQ each summer. Nisbett, of the University of Michigan, and his associates, elaborate on this finding in their 2012 book (on page 138). Several studies such as those reported by Herrnstein and Murray in 1994 (on page 405), and Currie in 2001 have provided extensive evidence (on pages 217 to 226 of their work) that preschool activities increase IQ scores. Additionally, Bronfenbrenner, in his work entitled *Is Early Intervention Effective?* published in 1999, reported that parenting training increases their children's IQ.

This, though only a partial list, is shown here to convey that environmental factors strongly influence intelligence, at least as measured by IQ tests. These observations are expanded in Chapter 4. Thus, while this should dispel any remaining myth that intelligence is innately determined, the lingering and persistent debate is, still, to what extent intelligent behavior is influenced by genetic factors as compared to environmental factors.

Questions about the measures of intelligence testing

Many questions may be asked about measures of intelligence such as,

- Why have girls had lower IQ scores than boys on spatial relations?
  a. Why is this changing in recent decades?
- Why are IQ vocabulary scores higher for girls than boys?
- Why are IQ math scores higher for boys than girls?
- Why are overall IQ scores very similar for boys and girls?
- Why do scores of white boys more frequently occupy the highest and lowest IQ spectrum than do white girls?

---

[20] Jencks and associates, (1972)

- Why do black girls occupy the highest IQ score spectrum more than do black boys?
- Why do blacks score 10 to 15 IQ points lower than whites?
- Why do the economically disadvantaged have lower IQ scores than the advantaged?

The answers to these questions do not point to genetic factors but environmental factors.

A projected new range of IQ

Certainly, a range of intelligence, as well as a range of IQ, will always exist among humans, but the range may be raised and narrowed. (The 95 percent population IQ range can be raised and narrowed from the present 70 to 130 to, say, 90 to 135.) As will be shown, the IQ scores of most people, currently with IQ scores at or below 87, can be enhanced so that their IQs move into the near average range and so that they become productive and self-confident people in society. The two-tiered society consisting of IQ haves and IQ have-nots, with all its potential conflicts, need never occur. As will be shown in Chapter 5, the average intelligence of blacks, as a group, need not be different from that of whites, as a group. Among whites, Jews have been shown, as a group, to have higher IQs than other groups of whites. And, here again, no evidence has ever been presented to suggest the reason is genetic; rather the reason is likely motivation and expectations. This may also explain why Asians have higher IQs than do whites. To reiterate, there is no evidence showing genetic factors explain differences in intelligent behavior between any groups of humans, whereas there is a great deal of evidence showing that environmental factors explain differences in intelligence between groups of individuals.

Other considerations that influence understanding intelligence

As stated, but worth emphasizing here, while *environmental factors do not alter genetic factors*[21], *environmental factors significantly influence genetic factors,*

---

[21] Notions of epigenetics, though relevant, are not discussed in this book.

for example, the effectiveness of the neuronal system. Such environmental limitations inhibit intelligent behavior in the moment, at all ages. And, because these constraints affect intelligence behavior in the moment, they also affect the effectiveness of the accumulation of information on which subsequent intelligent behavior is often based. Thus, environmental limitations impede not only intelligent behavior in the moment but also impede the gradual growth of an individual's intelligent behavior over time.

This basic notion—environmental limitations inhibit intelligence—is supported by many, probably most, researchers, including Richard Nisbett of University of Michigan, James Flynn of Duneeden University, Arthur Jensen who was of Stanford University, Ulric Neisser who some refer to as the father of cognitive psychology, Hans Eysenck at University of London Institute of Psychiatry, Richard Herrnstein who was of Harvard University, Richard Lynn of the University of Ulster, Charles Murray co-author of *The Bell Curve*, Sandra Scarr of the University of Virginia, Robert Sternberg of Yale University, and many more.

To the best of my knowledge, of the thousands of articles on this subject, all convey that nutrition and health have a significant effect on intelligent behavior. The question is not whether these and other environmental factors significantly influence intelligent behavior, but rather, by how much these factors influence intelligent behavior. Well-being influences intelligence both directly, determining the effectiveness of intelligence functions in the moment, and indirectly, determining the effectiveness of absorbing information from informal and formal learning experiences for subsequent use. One researcher, Martorell, explained in his 1998 work that *poor diet* and illness have the greatest effect on intelligent behavior during developmental periods. When the diet and well-being of children living in relative poverty are enhanced to adequate levels, their acquisition of age-appropriate knowledge, cognitive skills, and their scores on intelligence tests substantially improve.

As expanded in Chapters 4 and 5, because environmental improvements enhance intelligent behavior, especially of the disadvantaged, and because, therefore, education of the disadvantaged would prove more effective and likely more interesting, disadvantaged children would more likely remain in school through completion, be more capable, and subsequently, more employable. The *problems of society* would likely be significantly reduced.

Many believe these *societal problems* stem at least partially from family-cultural considerations. This book addresses these concerns some-what. For example, while it is known that preschool education enhances intelligent behavior and better prepares children, advantaged and disad-vantaged, for later schooling, it is also known that if parents participate in parenting training during their children's preschool training, there is a far stronger effect on their children's intelligent behavior and success in school. Incidentally, advantaged children currently do experience preschool educa-tion, good preschool education; the disadvantaged, in contrast, hardly expe-rience good preschool education, and rarely attend preschool at three years of age. (Why do we call pre-kindergarten education, preschool education, when it clearly is school education, indeed it is early education, and should be considered part of one's total education.)

Incidentally, many, if not most of the works referenced in this book are peer reviewed research, often replicated research, documented in the litera-ture, supporting the realization that health and nutrition significantly affect intelligent behavior and that nutritional and health constraints, all-too-of-ten, and usually more severely experienced by the disadvantaged, limit the development of one's intelligent behavior.

Less than widely known patterns of intelligence

A few additional patterns surface in reviewing the many studies. They indi-rectly affect the apparent influences on intelligence.

- Of the parents, the mother is the more significant force in deter-mining a child's intelligence. This becomes evident when reviewing several studies that show, for example, if the mother is the brighter of the parents, the child will likely be brighter than if the father is the brighter of the parents. This pattern may be changing and may change more so in the future, but, however much it changes, the pattern will likely remain because the mother will remain the more dominant nurturing force at least through prenatal care. This highlights, yet again, that the environment is a strong influence on intelligence.

- For offspring being a product of a mixed marriage, one parent black and one white, if the mother is white the child is likely to be brighter than if the mother is black. This has nothing to do with genetic factors because half the genes come from each of the parents. Nor does this have to do with race, because in both cases one of the parents is black. It only relates to the environment.

  The reason the child with the white mother is likely to be brighter than the child with a black mother is because the white mother is likely to have been more advantaged than the black mother and be able to provide the child with more advantages including, but not restricted to, more adequate prenatal nutrition and more adequate postnatal nutrition.

- No matter how significant one views genetic factors in determining an individual's intelligence, the significance of the environment in determining the intelligence of the offspring must be recognized. This becomes apparent when reviewing studies where the environments are similar for several different type offspring and where their IQ is essentially the same. As will be seen shortly in Chapter 4, the Smilansky studies show this, as well as the Eyferth studies discussed in Chapter 5, and, as will various adoption studies.

These observations highlight, again, the importance of environmental influence on intelligence.

Before we probe the known interventions that reduce the environmental inhibiters and constraints on intelligence – the known ways to enhance intelligence – most substantially of the disadvantaged, we more closely examine poverty and economic inequalities. The next chapter briefly discusses the frustrations, pain, and currently unnecessary hopelessness of the poor, as well as the hardly tapped economic opportunity, experienced by all, when the poor move into the middle-class increasing demand for goods and services largely provided by and benefiting the advantaged.

Myths have dominated the study of intelligence for generations – many centuries. For example, intelligence is not largely genetically determined, as many still believe. In fact, that part of intelligence that is genetically determined must be supported prenatally by the mother's adequate nutrition and good health or some genetic features do not develop, and postnatally these and other environmental factors must be provided without which the neuronal system is less than effective. Furthermore, intelligence grows with every formal and informal learning experience, again, depending on the well-being of the neuronal system, and these learning experiences are entirely environmentally determined. There is reason to believe that the real-estate of the neuronal system becomes increasingly occupied by these learning experiences.

Another myth is that intelligence is stable over time; surely, IQ scores are stable over time, *by design*, but, as hardly known, for IQ scores to remain stable, intelligence must grow. Because IQ scores appear stable, the general-public believes and even some professionals claim that intelligence is stable. Intelligence grows throughout life with every formal and informal learning experience; IQ scores are intended to be age independent, and therefore stable, whereas intelligence is age dependent – is not stable.

Another example of myths is the notion of *g* (general intelligence). This concept has become widely accepted among academics and has been with us for over a century. The data clearly shows that when someone is strong in one area of intelligent capability, like math, he or she will also be strong in other areas. No evidence has ever surfaced to support this claim genetically whereas a great deal of evidence has shown the cause of *g* to be environmental; strong nutrition and health, for example, determine *g*. Those that studied this issue have been largely hereditarians and thus oriented to seek a genetic answer to this issue. They have been seeking answers in all the wrong places. Specifically, well-being permits one who is strong in one area, to be strong in many. Well-being influences not only areas of intelligence but areas of the arts, sports, and human relationships – every area of human activity. The observation, more than a century old, coincidentally and unfortunately focused only on intelligence misinterpreting *g*'s meaning to relate only to intelligence, and misleading generations of researchers to follow. *g* should have been more broadly recognized is applying across all human activities. Well-being determines success in all areas of human activity not just areas of intelligence. g should stand for general well-being, not general intelligence. At least the field of psychology must be free of this and many other myths.

Is it true, as so many people believe, that the poor are disadvantaged because, as a group, they are born – actually conceived – with less potential intelligence than the advantaged? Differences in intelligent behavior among groups of people is widely recognized but not understood, more to the point, widely misunderstood. For example, while blacks have a lower average IQ than whites, some believe, without any supporting evidence, that that is due to genetic differences. A similar observation is made about average IQ differences between the disadvantaged and advantaged; that this difference is due to genetic differences, again, notably, without any supporting evidence. In stark contrast, abundant evidence shows that the difference in IQ between blacks and whites, as well as the disadvantaged and advantaged, is caused by inadequate nutrition, poor health, and other inhibiting environmental factors. See Chapter 5 for evidence.

Is it reasonable, as so many people have come to believe, that it is inevitable that we have a two-tier society consisting of the IQ haves and the IQ have-nots, or, more generally, the haves and the have-nots? Why do so many still believe that blacks, on average, are born to be less intelligent than are whites? Why have so many come to believe, recently, that whites are born to be less intelligent than are orientals? Why have so many come to hold these notions with firm and deeply-rooted confidence?

Potential intelligent behavior is surely influenced by genetic factors and all-too-often constrained by environmental factors. That is, while intelligence is influenced by genetic factors, it is environmentally constrained a great deal among the disadvantaged and supported a great deal among the advantaged. That suggests that the difference in intelligence between the disadvantaged and advantaged is the environmental constraints and environmental support influencing their intelligence. Examining this observation closely, one concludes that genetic influence is determined as Turkheimer and associates indicate, at about 10 to 20 percent, with environmental influence, positive or negative, determining the remaining 80 to 90 percent.

These and many other myths have delayed progress towards

understanding intelligence. They have caused the study of intelligence to remain mysterious, while understanding has significantly advanced in many other fields. We must develop an understanding of intelligence minimizing its myths, magic, and mystery. Chapter 7 analyzes and debunks a dozen such myths.

To what extent do genetic factors and environmental factors influence intelligence

Though many people seem to take great pride that they are unique, many differences are just that, just differences; our kidneys, lungs, heart, and blood are so much alike that they are interchangeable. On the other hand, surely, some of our genetic differences give some of us an advantage in sports, the arts, or in the sciences.

But this only considers part of the story. Genetically determined neuronal systems are influenced by nutrition, health, and other environmental factors; the combination determines the human range of mental and physical capabilities. Inadequate nutrition inhibits height, as we see among North Koreans compared to South Koreans. Inadequate nutrition also constrains intelligence, every kind of intelligence. Additionally, motivation and exposure often determine which, if any, potential talents are developed. We are unhealthy at least part of our lives, perhaps undernourished for at least short durations, or not exposed to experiences that support our genetic orientations—if Mozart had not had access to a clavier, we would not now enjoy the fruits of his creativity.

It is the combination – the interaction – of genetic and environmental influencers that determine one's physical and mental abilities. Most importantly, environmental influencers, such as nutrition, determine one's mental and physical capabilities far more than currently acknowledged. Try it yourself. Don't drink water for two days. You'll see, at least others will see, how incoherent you can be, no matter how brilliant you might otherwise be. On second thought, do not try this while reading this book. Actually, don't try this at all. Certainly, if Einstein had not had adequate nutrition, he would not have been nearly as creative. It is the blend of genetic factors and environmental factors that determines the likelihood of a Beethoven, a Madame Currie, a Jesse Owens, or an Einstein.

And then there are other differences that are not genetically or

46

environmentally determined. Every zebra has a unique set of stripes. Every giraffe has a unique set of spots. Every flower, even on the same plant, is unique. These differences result from nature being less than consistent, and that seems quite okay.

There are several ways to view these central comments and observations of the human condition. Some strongly believe that limited intelligence of the poor is largely genetically determined, and that currently not much can be done about it. In contrast, evidence overwhelming shows that inadequate nutrition, poor health, and other environmental constraints, typical of the disadvantaged, constrain their intelligence – preventing them from reaching their otherwise reasonably expected capabilities.

While the relationship between poverty and inadequate nutrition is widely known, and the relationship between nutrition and intelligence is surely known at least by academics and professionals, the direct relationship between poverty and intelligence is hardly known and all-too-rarely analyzed. Conveying an understanding of this relationship is the key motivation to have written this book. Currently, this relationship is hardly understood *partly* because psychologists have not yet developed a *definitive* understanding of the relative effect of genetic influences and environmental influences on intelligence, the cause of continuing great debate. Let's examine that for a moment.

Explicitly, one camp of professionals[22] has shown and firmly believes that genetic factors are more significant in determining intelligence than are environmental factors. This camp, therefore, concludes that the disadvantaged are less intelligent than the advantaged largely because of genetic reasons about which, presently they feel, not much can be done. The evidence for this conclusion includes research with children across the entire spectrum of socio-economic status.

Another camp of professionals[23] has shown and holds a significantly different view; that, especially among the disadvantaged, environmental factors are far more significant than genetic factors in determining one's intelligence. This position is validated by studies that include only children

---

[22] For example, Herrnstein and Murray, 1994; Jensen, 1969 & 1998
[23] For example, Turkheimer and associates, 2003

that are disadvantaged – children belonging only to the lower portions of the socio-economic spectrum.

While there is continuing debate between these two camps, importantly, there is a much larger third group which includes most professionals of both these camps. These professionals essentially agree that environmental constraints, common among the poor, substantially constrain intelligence, regardless of underlying genetic factors. Assuring adequate nutrition, good health, less stress, adequate motivation and expectations, and therefore more effective formal and informal learning experiences significantly increases the intelligence and capability of the poor and thus opens their path of upward mobility – increasing the middle-class and decreasing poverty.

Evidence overwhelmingly indicates that inadequate nutrition and poor health prevent the disadvantaged from reaching as close to their potential intelligence as is far more closely reached by the advantaged. Strong evidence supports the notion that the single most important cause of poverty is constrained intelligence. This limits capability, opportunity, employability, upward mobility, and therefore, not only causes poverty but causes poverty to persist across generations. This book shows that this circular pattern – poverty causes poverty – can be broken.

Surely, some will correctly and rationally believe that people like Mozart, Einstein, and many others were born with more, perhaps only different, genetically determined, benefits than others. While that would be difficult to refute, it would not diminish the power of nutrition, health, orientation, and other environmental factors. If those strongly endowed people had experienced limited nutrition and poor health during their developmental stages, their genetic gifts would have been limited, perhaps lost. It is probably true that millions of such people are born into poverty every year and never leave poverty, their potential being lost or, squandered.

CHAPTER 3
Poverty

I have a religion. You may call it blasphemy. It is that there is a god for the rich man and none for the poor.[24]

Mark Twain

Why are there so many poor people? Why are children who live in poverty, likely to perform relatively poorly in school? Why do many of these children remain in poverty for much or even all their lives? Why do the poor feel little potential to work their way out of poverty? Why do the advantaged assume not much can be done to reduce poverty? These are tough questions that have frustrated societies for ages. Recently, some progress has been made towards reducing worldwide poverty but much more could be done, and much more quickly, if its most important underlying cause, as described in this book – decreasing constraints on intelligence, in effect enhancing intelligence, especially of the disadvantaged – was more directly addressed.

Some economists, such as Joseph Stiglitz, a Nobel Prize winner in economics, support tax modification approaches to solve the economic inequalities and opportunity inequalities. Stiglitz's recent white paper indicates that a fair tax structure would go a long way to resolve these concerns. He addresses the broad economic considerations, such as improving infrastructure (roads, bridges, airports...), reducing tax avoidance mechanisms, and decreasing unemployment by returning jobs to the United States. For example, Stiglitz's

---

[24] Twain, M. (1869). *Innocents Abroad or the New Pilgrim's progress*

paper points out that currently corporations are motivated to leave profits abroad, which when used for development creates jobs abroad.

While there is merit in these, just mentioned, approaches, they are *not* part of this book's agenda. Interestingly, economists hardly mention enhancing intelligence to reduce economic inequalities, or, inequalities in opportunity. That is not their orientation; notably it is not the orientation of any discipline. But *enhancing intelligence is the central orientation, the heart, of this book.* Here, we focus on a set of related approaches and solutions that, collectively, and directly, decrease environmental constraints on intelligence thus increasing intelligence and capability and thereby, helping the poor achieve closer to their potential. Enhancing intelligence improves the likelihood of the less fortunate to experience the upward mobility path and helps them develop more self-respect and a sense of ownership in society. These comments are expanded in the following chapters.

A brief discussion of economic inequalities in the United States

To a significant degree, children in poverty often lead a relatively hopeless life. Not only are they inadequately nourished and have relatively poor health, but they live in dangerous neighborhoods, and experience much greater stress than do advantaged children. These problems are much more severe in the developing countries, where, in addition, the poor live, for example, with severe sanitation problems.

Towards achieving an awareness of the extent of the problems being considered here, one might want to know that in the United States, 25 percent of children live in poverty. About 18 percent of the total population lives below the poverty line, on less than about $28,000 per year for a family of four. Additionally, a definition for living in *near* poverty,[25] and accordingly qualifying to get free or reduced-priced lunch for children in the United States, is an annual family income of about $56,000. Notably, the median[26]

---

[25] The National Center for Child Poverty defines "poor" households as those with annual incomes that fall within the HHS poverty guidelines and "low income" households as those with incomes less than twice the amount set by guidelines (Cauthen and Fass, 2008).
[26] Medians often present a more representative picture than do averages; while the median income in the United States is about $52,000, the average income is considerably higher because it is weighted by those incomes at the high end of the spectrum.

family annual income in the United States is about $52,000, lower than the near-poverty line. Thus, though hardly acknowledged, most people in the United States live in poverty or in near-poverty. Startling!

In contrast, wealth in the United States is disproportionately concentrated in the hands of a very few; the four hundred richest people are worth $2 trillion ($2,000,000,000,000). That statistic alone does not suggest a cause of the prevalence of poverty however the data conveying that 95 percent of the growth of the economy in the United States goes to 1 percent of the population does suggest a cause of the widespread existence of poverty in the United States; the income of the working class has essentially remained stable for the last several decades. Thus, while most of the population in the United States lives in poverty or near-poverty, and a relatively small percentage of the people live in comfort, a miniscule part of the population lives in extreme wealth. That, it seems, defines a two-tier society, a set of *gated* communities, with all its real and implied problems and conflicts.

Joseph Stiglitz has recently written that America has among the worst economic inequalities of the developed countries. He wrote,

> The gross domestic product of the United States has more than quadrupled in the last 40 years and nearly doubled in the last 25, but as is now well known, the benefits have gone to the top—and increasingly to the very, very top.

What might explain this? On January 23, 2015, Mark Shields, a political columnist, commented on the PBS NewsHour,

> Between 1948 and 1973, the productivity per hour, that is the goods and services produced by the average American worker, went up 96 percent, and their wages went up 91 percent. It was the golden era. In the forty years after 1973 productivity, again of the workers, went up some 76 percent and at the same time their income, wages, went up only 9 percent. We have a mal distribution of wealth.

Notice, please, this represents an imbalance in the sharing of the fruits of production; unlike the previous twenty-five years, productivity gains in the most recent decades were mostly enjoyed by employers, hardly by workers. This is not a question of distribution of wealth. It is a question of recognizing the contribution of labor in the production process. That is a real concern of current economic inequalities. Of course, if sharing of production gains is unreasonable, then subsequently redistribution of wealth may need to be considered.

A week later, January 30, 2015, Larry Summers conveyed this problem differently. He said,

> If the income distribution in the United States were the same as it was in 1979, there would be a trillion dollars more in the hands of the bottom 80% of the population, or 11 thousand dollars per family, and a trillion dollars less in the hands of the top 1%. ... So there has been a big redistribution that has worked to the detriment of the middle-class. And unfortunately, at the same time, we've made that worse for the middle-class with a set of changes that have come in; we have less progressive taxation than we did then, we are investing much less in the public sector than we did then, we have much less in the way of protections that strengthen worker rights,... so there have been a variety of changes that go beyond the financial crises, that have exacerbated the inequality and exacerbated the lack of growth.

Additionally, Stiglitz remarks,

> Last year, the top 1 percent of Americans took home 22 percent of the nation's income; the top 0.1 percent, 11 percent. [As reported by syndicated columnists Mark Shields, 10 percent of the population took home 50 percent of the nation's income.] Recently released census figures show that median income in America hasn't budged in almost a quarter-century.

Economic upward mobility is a measure of a society's well-being. However, the bottom rungs of the ladder have become sticky, says

Washington Post columnist, Michael Gerson. A question one may ask is, why? Improving upward mobility is a major objective of this book.

While the focus of this book is to dramatically reduce poverty by increasing the intelligence of the disadvantaged, economic inequalities severely impact poverty. Worse yet, the growth of economic inequalities in the last 35 years is deepening poverty. We are occasionally reminded that in the United States the richest 1 percent own 37 percent of society's wealth while the poorest 60 percent owns only1.7 percent. Bernie Sanders, Independent Senator of Vermont, adds that when one family, the Walton family, has more wealth than the bottom 40 percent of the people of the United States, a problem should be apparent. Amazingly, the people have heard these data, yet continue to support those forces that tend to advance economic inequalities—a strange dilemma. But maybe not so strange after all; perhaps the individuals who are poor or near-poor feel that one day they might be one of *them*, and that the dichotomy, then, would feel okay—a stunning indictment on human nature. Of course, the real problem remains, that being that more than half the people of the United States are disadvantaged, considered either poor or near-poor.

A CNN poll, reported by Tami Lubhy in June 2014, indicates that "59% feel the American dream has become impossible to achieve"; the American dream is out of reach. However, optimistically, Sanders says, get people together who may not agree on every issue, but they understand that gradually the middle-class is collapsing and that we are moving toward an oligarchic form of society where billionaires run the country. He suggests that when the people lead, the leaders will follow, and this author would add, for good reason.

Though these concerns have become widely recognized, solutions remain narrow and focused. People representing the spectrum of ideologies, not only prevalent in the United States but pervasive across the planet, have suggested solutions from, "the poor should pick themselves up by their bootstraps", to, "we need a redistribution of wealth". Neither of these extreme solutions appears to address the core problems directly. We are dealing with complex social problems. Institutions need to change. Values need to change. Consideration must be given basic values including the value of each person, the value of *people*, the value of their *potential*, the cost to each poor individual and the cost to society of wasting their enormous collective potential, and the value of their *diversity*—societal, racial, cultural, and in other ways that humans across the planet differ.

Let's digress for a moment to bring some of these smoke-and-mirror notions down to earth. As a disadvantaged child becomes *more* adequately nourished, he or she can more effectively focus and attend to formal and informal learning experiences, can advance in school at a more appropriate rate, may more likely find more interest in school, will more likely graduate, be more employable, less prone to criminality, more likely to live in reasonable comfort, feel pride in himself or herself, and feel more human dignity. He or she will value himself or herself more, and so will others.

Let's view this discussion from a different perspective, one that surfaces the profound human sensitive issues. Having volunteered to mentor and tutor disadvantaged children in local school systems, I worked, through the years, with dozens of children, always one-on-one, following them as they moved from school to school. Most of those children were labeled learning-disabled, but it quickly became evident that the disability of many was the baggage they carried – their family baggage. One child lived with his grandmother. His mother was divorced and lived with her partner, essentially ignoring her son. This child told me that he had seen his mother on his birthday. "When was that?" I asked, and he said, "about a half year ago". I then asked, "How did that go", and he told me that she came with a gift but had to leave immediately because her friend was waiting in the car. One child's mother was thirteen years older than him. I listened to these children, and learned, among other things, that their potential had hardly been tapped. Their stories are sprinkled throughout this book to provide real meaning and sensitivity to the data. This book is the result of their stories and years of research and analysis. And while their stories do not reflect economic issues directly, they do deal with core concerns that inhibit people from getting as close as they could to their potential.

This anecdotal background, with its generic and underlying societal issues presents a platform upon which fundamental questions may be asked, scientific data may be analyzed, and rational conclusions may be reached, and upon which public policy may be appropriately questioned and rational policy may be based.

Much has been written about these problems, but articulating the concerns is not enough. What is it that the poor in the world have in common besides poverty? *They have vast, essentially untapped, human potential;* their capabilities, talents, and intelligence are hardly realized – largely wasted. The purpose of this book is to identify underlying solutions to this problem; to reduce constraints on opportunities such that poverty, for most, is eliminated. The intent is to also surface the awareness of the enormous profits and benefits left on the table by the advantaged. Again, while *poverty has always been with us, and it is not likely to be entirely eliminated, it need not continue to be so widespread?*

Problems of poverty in the developing world are, surely, worse than those in the developed world, yet solutions are quite similar

This book deals indirectly with economic inequalities which have been growing in the last three decades. Significantly, Angel Gurria, Secretary-General of the Organization for Economic Cooperation and Development (OECD), writes, "for every 1 percent that inequality grows [there is] a drop in economic growth of about 0.2 [to] 0.3 percent. That means a more unequal society will grow less and inequality becomes an obstacle to growth in and of itself." Stated positively, the more economically balanced a society becomes the more economic growth. She adds that three decades ago the top 1 percent took home 10 percent of national income whereas now they take home over 20 percent. This presents a formidable obstacle for economic growth. More to the point and key to understanding this set of problems, as inequalities diminish life will likely improve for all.

Moreover, Gurria adds, "The problem is not just a question of income. It is also a question of inequality of opportunity, inequality of access to health services, inequality of access to education, inequality of access to employment opportunities. And these [inequalities] have been also growing."

Expanding on these problems a bit more, while about a half the population living in developed countries, at least in the United States, lives in poverty or near-poverty, as may be expected poverty is far worse in the developing and underdeveloped world. As reported by the World Bank in 2008, before the Great Recession, nearly half the world's population lived on less than three dollars per day, of which about 2.4 billion people lived in severe poverty on less than two dollars per day, of which 1.4 billion lived in extreme poverty, on less than $1.25 per day, and many of those in Sub Saharan Africa and parts of Latin America lived on less than seventy-five cents per day.

Progress, however, is being made, reported the *Economist* on June 1, 2013.

> In 1990, 43% of the population of developing countries lived in extreme poverty (then defined as subsisting on $1 a day); the absolute number was 1.9 billion people. ... By 2010 it was 21% (or 1.2 billion; the [extreme] poverty line was then $1.25 ...) If 21% was possible in 2010, [they write] why not 1% in 2030?

While this shows a significant reduction in poverty has occurred for those living on less than $1.25 per day, how about all those living on less than $2.00 per day, or $2.50 per day, or in the other levels and forms of poverty? It does not look good as shown in figure 2.

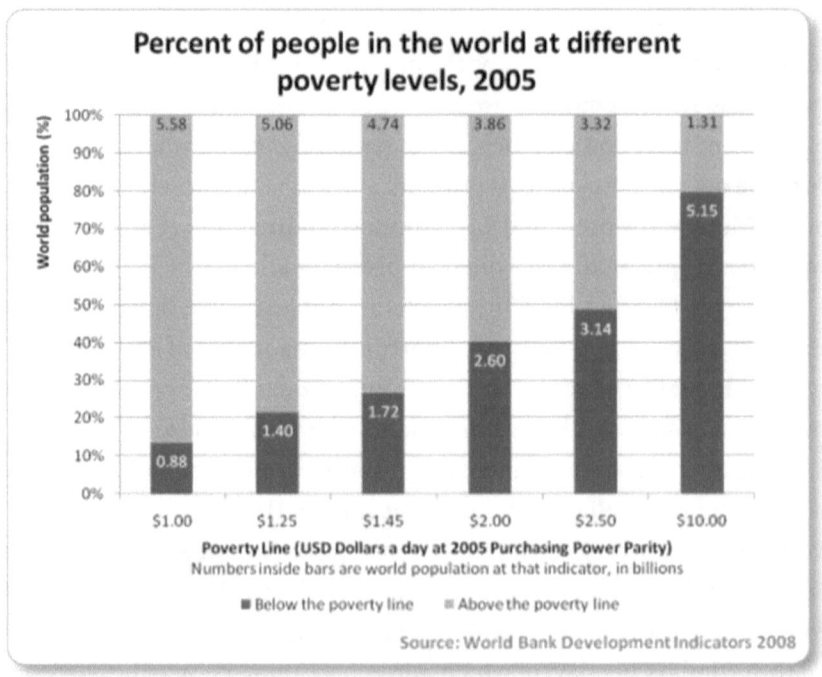

Figure 2

Figure 2 shows that, for example, 2.6 billion people, 40 percent of the world population, lived on less than $2.00 per day in 2005; that 3.14 billion people, nearly 50 percent of the world population, lived on less than $2.50 a day in 2005.

Furthermore, the Economist's data mostly reflect changes in China (See Figure 3). The *Economist* indicates, "China, alone, accounts for around three quarters of the world's total decline in extreme poverty over the past 30 years." Most of this reduction in extreme poverty was the result of dramatic growth in GDP. And while this economic growth results in benefits to the poor, it has reduced poverty much less than might have been expected.

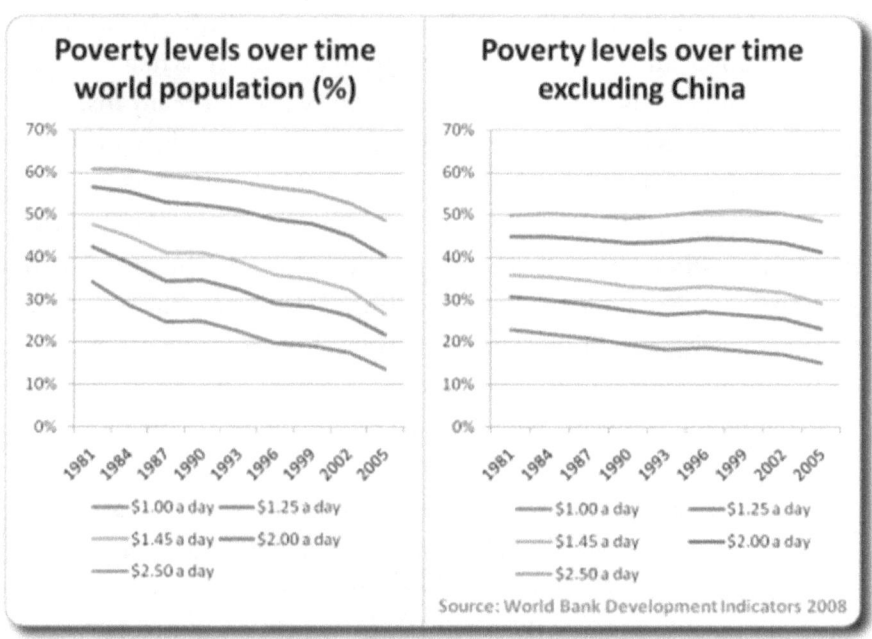

Figure 3

The left and right part of Figure 3 shows that just under 50 percent of the world population lived on less than $2.50 per day (the top line) in 2005; that 40 percent of the world population lived on $2.00 a day (the line just below the top line) in 2005. The bottom line on the charts show that about 15 percent of the world population lived on less than $1.00 a day. Interestingly, the left side of figure 3 shows the decline in poverty of the world population including China, whereas the right side of figure 3 shows the decline in poverty of the world population excluding China, a marked difference.

By way of clarification, the *Economist* conveys that both growth and income distribution matter. They quote Martin Ravallion, until recently the World Bank's head of research, who "reckons that a 1% increase in incomes cut poverty by 0.6% in the most unequal countries but by 4.3% in the most equal ones"—the more equality, the more poverty decreases. This is additional support for the report by Gurria mentioned just above. It suggests that reducing economic inequality along with increasing GDP is effective in reducing poverty and that the more the reduction in economic inequality the more effective the reduction in poverty. This reveals an upward economic spiral benefiting all. This book conveys ways to increase intelligence, thereby improving opportunities, increasing employability, and thus reducing poverty. Accordingly, this paves the upward mobility path for many disadvantaged to move into the middle-class, and it increases GDP, benefitting all, including the advantaged.

More recently, Christoph Lakner, a consultant at the World Bank[27], declared that while national economic inequalities may be growing, international economic inequalities are falling. This reflects the egalitarian movement in China and India but, when averaged over the planet, presents a misleading perspective.

The *Economist* conveys an additional concern.

> The changing geography of poverty will pose different aid problems over the next 20 years. According to Mr. Chandy [Ledlie, and Penciakova, as reported in a study for the Brookings Institution, a think-tank in Washington, DC], by 2030 nearly two-thirds of the world's poor will be living in states now deemed "fragile" (like the Congo and Somalia). Much of the rest [of the poor] will be in middle-income countries. This poses a double dilemma for donors: middle-income countries do not really need aid,

---

[27] Reported in the *New York Times* by Cowen, T. Income Inequality Is Not Rising Globally. It's Falling; Economic View, July 19, 2014

while fragile states cannot use it properly. A dramatic fall
in [extreme] poverty requires rethinking official assistance.

This book addresses these considerations not through foreign aid, as we know it today, but rather, specifically, by supporting the means for upward mobility; to provide interventions to enhance intelligence so that the less fortunate may more effectively help themselves, in both the fragile and middle-income countries.

In his book, *The Myth of Development; Non-Viable Economies and the Crises of Civilization*, Oswaldo de Rivero, a former Peruvian ambassador to the United Nations, expands on these concerns conveying that most of the developing countries are hardly developing; the fragile states, he observes, will likely remain fragile. He explains this by relating that the benefits of globalization are mostly going to the developed world, while little is going to the developing world. Similarly, technological advances are largely economically benefiting the developed countries. This is because research and design of technological innovation are essentially developed in the advanced countries. Also, technological products are increasingly being produced in developed countries near the largest markets because technological advances, such as robotics, permit products to be built closer to the market with much less, though skilled, labor and therefore, less need to seek the cheap labor abundant in the developing countries. Moreover, developing countries do not yet have technologically skilled or well enough educated people to become technologically educated. Furthermore, developing countries can no longer find enough profits in providing just resources such as minerals because, based on technological advances, raw materials are being replaced by synthetic materials that are often stronger and less costly.

In parallel, de Rivero writes that the problems of the underdeveloped countries stem partly from their continuing high population growth and their rising urbanization which, in turn, yields increased overcrowding with enormous problems in the supply of water, food, and energy. Incidentally, some population growth has been shown to be good for growth of GDP, but high population growth is a basic factor detrimental to growth. Also, because poverty is so widespread in the underdeveloped countries, and because nutrition is less than adequate, education, whatever it may be, is correspondingly less than effective, and the population does not become well enough educated to become a technologically fit society. With the growth in population, water per person becomes even scarcer, and, because water becomes less plentiful, food becomes less available and more costly. With the increase in population comes an increase in need for energy which in-turn reduces available funds for education, food, and sanitation. De Rivero concludes that unless the current pattern is changed dramatically, many of the so-called developing countries will not likely become viable.

In contrast, some developing countries such as China and India have recently made significant progress toward reducing poverty, though poverty still overwhelms their societies. Though three hundred million people in China have entered the middle-class in the last two decades, eight hundred million still live in various levels of poverty. And importantly, a cost of moving out of poverty in China has been a dramatic increase in environmental destruction. Air pollution due to coal-fired energy plants has resulted in about one million cancer-related deaths in China. China is experiencing an increasingly serious shortage of water, especially in the northern regions. Though not as severe in China because of its previous one child per family policy, population increases and increasing urbanization in other developing countries has resulted in extreme overcrowding, severe sanitation problems, increasing water shortages, food shortages, and delinquency.

Stiglitz writes,

> For these reasons, I [Stiglitz] see us entering a world divided not just between the haves and have-nots, but also between those countries that do nothing about it, and those that do. Some countries will be successful in creating shared prosperity — the only kind of prosperity that I [he says] believe is truly sustainable. Others will let inequality run amok. In these divided societies, the rich will hunker in gated communities, almost completely separated from the poor, whose lives will be almost unfathomable to them, and vice versa. [Stiglitz writes] I've visited societies that seem to have chosen this path. They are not places in which most of us would want to live, whether in their cloistered enclaves or their desperate shantytowns.

I remember strolling with my wife on a pristine beach in the Caribbean when we saw, about two hundred feet from the beach, a large estate surrounded by a wall topped with glass shards and electrified wire. We could not see into the estate and they, it seemed, could not see out. We discussed the dichotomy and, as we continued down the beach, decided we were much better off. We were enjoying the magnificence of nature, and they were suffering from apparent insecurity from within what appeared to us as useless wealth.

Are the possible solutions to reduce poverty, economic inequalities, and inequalities of opportunity in the United States also applicable for the comparable international concerns? Yes, as has been shown, and will continue to be shown in this book.

Based on more recent studies, as will be seen within the next two chapters, while intelligence is surely partly influenced by genetic factors, it is greatly influenced by environmental factors. Inadequate nutrition among the disadvantaged, and, adequate nutrition among the advantaged strongly influence intelligence. As is expanded in Chapter 4, among the disadvantaged intelligence is largely determined, 80 to 90 percent, by such environmental factors as limited nutrition. This was shown by Turkheimer and his associates in 2003. Thus, heritability, at least among the disadvantaged, is 10 to 20 percent. Arguably, this is also true for the advantaged. This is clarified early in Chapter 7.

Harvard political scientist Robert D. Putnam has written several books on related subjects. His latest book, *Our Kids: The American Dream in Crises,* deals with concerns of economic inequalities. His book, like this one, focuses on inequalities of opportunity. He conveys that early childhood development as well as other sociological activities of our young needs improvement. Researchers suggest different interventions to resolve various issues of society. A common, though usually unsaid, thread underlying these interventions is *enhancing intelligence.* It is the clear and primary focus in this book. With intelligence enhancement, achievable in many ways, especially among the disadvantaged, these problems are greatly diminished.

One more inequality, the focus of this book, is *intelligence* inequalities which, it is asserted, is a major cause of economic inequalities, opportunity inequalities, poverty, and other societal problems. *Discussing this set of problems from the viewpoint of intelligence inequalities makes clear that we are not suggesting taking from the rich to give to the poor; one cannot take intelligence from one person to increase the intelligence of another.* And we know many environmental interventions that would dramatically increase intelligence thereby reduce intelligence inequalities, and, in-turn, reduce economic inequalities, opportunity inequalities, and poverty.

Specifically, while many different interventions are promoted, in this book and elsewhere, this book focuses not only on each intervention but on the collection of interventions to achieve a single objective – to enhance intelligence, to decrease poverty and decrease intelligence inequalities. As has been mentioned, the advantaged behave more intelligently, as measured by IQ tests, than do the disadvantaged. The good news for all of society is that one can increase the intelligence of all of us, most substantially of the disadvantaged, thereby reducing intelligence inequalities.

Opportunity

Poverty often results from limited opportunity. Equal opportunity for all cannot be achieved, however opportunity inequalities may reasonably be minimized. Occasionally opportunity is restricted by errors in judgment. Digressing for a moment to place this in perspective, some constraints of opportunity appear inconspicuous to some but all-too-often, to the individual, are powerful and perhaps demoralizing.

> As an example of a demoralizing situation, one sensitive and lively girl, let's call her Dottie, who I had tutored and mentored for a couple of years and who had been labeled learning-disabled, had not been treated as learning-disabled for over a year; she happily and appropriately had been enrolled in regular classes. At the beginning of a school year, Dottie's mother called me to say that her daughter had just come home crying and terribly upset because she had been placed in a resource class, a class for learning-disabled children. Why did that happen? Would I please look into it?
>
> As it turns out, during the summer months a teacher for learning-disabled children had been hired by the principal. The new teacher easily determined which children had been in classes for learning-disabled students the previous year. But then she noticed

that Dottie, who was labeled, learning-disabled, had not received services the year before. The new teacher assigned Dottie to her class. Hearing all of that from the guidance counselor and knowing that Dottie had been functioning very well in regular classes, *and*, knowing that the school cannot, legally, merely place a child into resource classes without the prior consent of the parents, I suggested that she be returned to regular classes because apparently an error had been made.

And here is where the injustices started. I was told by the guidance counselor that she will investigate it and let me know within a few weeks. I told her that Dottie was very upset, embarrassed in front of her friends, and that this needed to be corrected that day. Upon being told to be patient, I got up and said that I was going to the principal now. Recognizing my advocacy role, I was asked to wait a couple of days. I did, and then was told a meeting was arranged between the assistant principal and several others involved three weeks from then. I responded that a decision was needed that week. It was, and the child was returned to regular classes. Not only did she do well, but she, unsolicited, formed a group of children to tutor children in lower classes who needed help.

Without advocacy, children are often less than protected, and, in this situation, there is no telling what might have resulted. Not so surprisingly, the system often protects itself first. The students are claimed to be of primary concern, but processes do not always work that way. Resource classes are often a necessary and good answer for children who need special help, but for those who do not need this help resource class will not only delay the child's normal progress in school, but also, as we have seen just above, turn the child against schooling and thus severely limit her opportunities in life.

These kinds of injustices may appear trivial to many people, but to this child it was severe, demeaning, and without an advocate it may well have been at least distressing if not devastating. Incidentally, most children do not have an advocate; parents usually feel the school personnel must know what they are doing, and that they must follow the school's direction. I must quickly add, school systems often do know what's to be done, but not always;

school systems are run by people, human beings, who occasionally make mistakes, and sometimes have biases, occasionally resulting in injustices that may best be appropriately and rationally surfaced immediately. Examples of good people being inhibited from coming close to their potential by well-meaning professionals is not uncommon and often has a lasting effect on their lives and that of their yet to be formed families.

Surely, the value of every child is immeasurable, especially to his or her parents, family, and friends. And while this value, an emotional value, is recognized by many—those in power as well as those without power—it is not the kind of value about which we are talking. Here, we focus on the value of a child's potential—an individual human being's potential—his or her self-satisfaction. For the disadvantaged around the world, including many in the developed world, this depends on opportunity, and importantly, it depends on reducing constraints on opportunity.

To be clear, again, we believe the problem of upward mobility is not so much, *economic inequality* as it is *opportunity inequality*. The solution is not simply redistribution of wealth—highly inflammatory terminology—but rather reducing constraints on opportunity, reducing the opportunity gap. Few would care, this author would not care, that the rich got richer so long as the poor left poverty. Talent of, and risks taken by, the rich should be rewarded. Of course, this must be true for the less advantaged as well. The development of talent should be encouraged. To achieve a more equitable distribution of opportunity, restrictions on developing talent, especially among those living in poverty, should be reduced; constraints on adequate nutrition and thus effective formal and informal learning experiences, which limit opportunity of the poor, should be reduced.

Having given this discussion a human context, we return to the broader conversation. While enormous productivity gains and economic growth have been experienced through technological gains, *human capability, more to the point the potential* of the disadvantaged masses, remains largely wasted. This can hardly be justified but it characterizes the history of humankind.

Over 50 percent of the population in the United States and the Western world is under-employed; even for many of those who are employed, their potential capabilities are scarcely tapped. While 80 percent of people on the planet, live in some level of poverty, these people and many more are under-employed; their potential talents remain hardly challenged. With enhanced human intelligent behavior, especially of the billions of disadvantaged, overall productivity gains will be even further improved through advances in *human* productivity, providing an additional and perhaps more significant spurt in global economic growth. The currently disadvantaged may more readily move into the middle-class. They may increasingly become technologically informed, will likely contribute substantially to GDP growth—and encouraging a wider share of the benefits— will provide the foundation for a greatly expanded world economy, and central to the objectives of this book, societies will likely build on a more equitable platform.

With the growth of only technological advances, humanity is quickly consuming global resources. However, if the still hardly tapped human capability is simultaneously, in parallel, advanced, then the global resource limitation problems are more likely resolved because more people, more capable people, could be creatively working on, and more likely solving, the many known and yet unknown problems.

On a related topic, we live in the most rapidly changing of times in all human history. Technological innovation and globalization have profoundly transformed life across the planet; medical advances, while initially benefiting those in the developed world, gradually spread to the developing world. However, while technological advances and globalization benefit many people in a lot of ways, they have exacerbated economic inequalities across the planet. Life remains rather grim for most people in the developing world and many even in the developed world. Importantly, as robotics, computer technologies, and information processing advance, more people at all levels will experience greater difficulty finding work. Surely, new job categories will surface, but new capabilities will likely be needed. The poor with limited capabilities and opportunities will likely carry the greatest burden of these advances. This can be reversed by enhancing the intelligence, capability, and adaptability of the poor.

As an interesting aside, one creative group of entrepreneurs, in the coal producing area of Kentucky, recognized that coal production will become decreasingly viable as compared to gas and other coal alternatives. They also recognized that coal workers do not want to leave their families, their environments, and their culture. So, this entrepreneurial group is now training coal workers to program computers in the mountains of Kentucky, near their families, making use of fiber optic connections to their markets. *History teaches us that well-understood problems seem to invent their own solutions.*

As probably noticed, this book conveys many interrelated notions, each represented by one or more of the many threads of a tapestry, threads of different colors and different textures. Some readers will be interested in some threads but not in others. However, collectively, the threads form a tapestry of society that reveals images and scenarios that need recognition and suggest change. Some will view the tapestry as too complex, too overwhelming, but when viewed in its entirety, solutions become apparent. The beauty of the tapestry when seen in its entirety is, in fact, its simplicity. While the tapestry depicts world problems, it also conveys local and even individual human issues. Although the tapestry shows injustices, it also shows opportunities. Though the tapestry conveys societal weaknesses, it also makes solutions apparent.

CHAPTER 4

IQ gap between the advantaged and disadvantaged, regardless of race, is entirely explained by environmental factors, thus may be essentially eliminated.

In this chapter we discuss several additional threads of the book's total tapestry. Each of these threads provides awareness of ways for an individual to reach closer to his or her potential intelligence. Collectively, these threads show how enhanced intelligence, especially and most substantially of the disadvantaged, would gradually lift most out of poverty, largely eliminating this plague of society. This chapter conveys research studies that support comments made earlier and is included to provide fundamental evidence to justify conclusions.

Charles Darwin's message mentioned earlier in this book, suggests that society implement ways to reduce environmental constraints where they exist but especially on the disadvantaged so that the poor may more closely reach their potential intelligence and accordingly more closely reach their potential capability. In this chapter we discuss differences in intelligence between individuals—not differences in intelligence between groups of individuals. To repeat for emphasis, there is no evidence that the disadvantaged, as a group, have lower intelligence than the advantaged because of genetic differences. Yet, there is a great body of evidence demonstrating that the disadvantaged, as a group, have lower intelligence than the advantaged because of environmental constraints, and that these may be corrected. Environmental ways to decrease group differences in intelligence are discussed here and in Chapter 5.

In this chapter we discuss environmental factors that influence differences in intelligence between individuals focusing on differences in well-being. Racial genetic differences, as will be seen, do not explain differences in intelligence. The next chapter, Chapter 5, quantifies the effect of these considerations. However, we first dispel several notions that have delayed a definitive understanding of intelligence, for generations.

Influence on intelligence of environmental factors and genetic factors

The genetic influences and environmental influences on intelligence are indistinguishable as soon as these influences are fused, merged, and become recognized as intelligent behavior. Clarifying this observation, environmental factors permit the expression of genetic factors; with limited nutrition genetic functions perform less effectively. After the genetic and environmental influences merge to become intelligent behavior, there is no way to know whether genetic or environmental influence was more significant in determining the behavior. Some might claim that this ambiguity can be resolved on a population basis, but here, too, after the environmental influences and genetic influences have fused, currently, there is no way to know whether genetic or environmental influence predominate. See Figure 4.

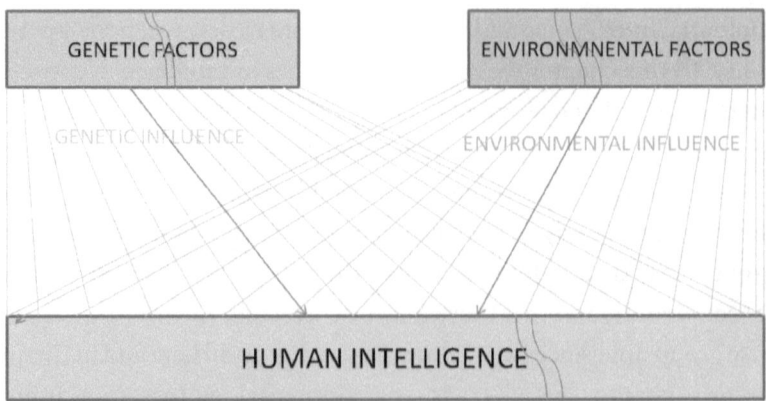

Ambiguity of Genetic and Environmental Influence on Intelligence

Figure 4. Ambiguity of genetic and environmental influence on intelligence

This ambiguity is further confounded because while we know the effect of environmental factors on intelligence, we do not know how to isolate the influence of genetic factors on intelligence. Specifically, we know the environmental effect on intelligence as, for example, when mineral and vitamin supplements are provided an individual, but, currently, we have no way to tell how one or another genetic factor, or even a combination of genetic factors effect intelligence. Moreover, we know intelligence grows with learning experiences, therefore environmental influences gradually grow over the life of an individual. Rationally, this causes the ratio of genetic and environmental influences to steadily decrease throughout much of one's life[28]; environmental influence increasingly dominates genetic influence. Yet, while the contribution of genetic influence and environmental influence is ambiguous and once fused are currently indistinguishable, this observation is hardly acknowledged. This author claims that the specific ratio of genetic and environmental influence on the mix is currently unknowable, and that, currently, with available scientific techniques the components of the mix cannot be differentiated. Nonetheless, many researchers assume the mix to be largely genetic, in a ratio of 80:20, while others assume it to be largely environmental, in a ratio of 20:80, the inverse; and they hold firm to their convictions rarely addressing this basic ambiguity; their positions are often conveyed with confidence and near certainty.

Researchers have collectively spent thousands of years of effort to understand intelligence yet hardly addressing this ambiguity directly. And this unanswered question only opens the discussion of at least two stubborn issues, namely,

- In what ratio do genetic factors and environmental factors influence an *individual's* intelligence?

The data clearly indicates different IQs for blacks and whites at age four. As shown in the next chapter, Martorell, in 1995, showed that those differences start developing at birth or even before.

---

[28] Interestingly, some researchers have provided evidence that shows that heritability appears to increase as an individual matures.

- Similarly, while researched by many, we still have no consensus on what causes differences in intelligent behavior between *groups of individuals, blacks and whites, or disadvantaged and advantaged.*

Heritability, at least among the disadvantaged, has been shown to be 10 to 20 percent. Furthermore, as will be shown immediately, this is, arguable, also true for the advantaged. These issues are addressed now as well as summarized in Chapter 7, "The Mythology, Magic, and Mystery of Intelligence".

Ratio of environmental and genetic factors influencing a person's intelligence

Heritability studies, using different methods and based on dozens of twin studies as well as studies with individuals of differing biological relationships, indicate that variation in intelligent behavior within a population is significantly determined by genetic variation within that population. This has been observed and reported by, for example, Bouchard and McGue, in 1981, and Scarr, in 1998. These studies have resulted in the understanding that genetic factors and environmental factors influence the intelligent behavior of individuals according to a ratio. The ratio of genetic to environmental factors is variously estimated at 80:20 (80 percent genetic factors to 20 percent environmental factors) documented by Jensen in 1969 and 1998, and 60:40 used by Herrnstein and Murray in 1994. However, there can be no specific ratio, universal ratio, across all populations because every population varies in environmental constraints. The specific ratio of genetic and environmental influence remains unresolved.[29]

These heritability estimates have been thought to be applicable to individuals from all ethnic backgrounds and socioeconomic statuses. However, those just mentioned were essentially derived from studying children of families across the entire spectrum of SES (socio-economic status). In 2003, Turkheimer and his associates showed that the ratio for disadvantaged children is essentially inverted, with environmental factors being much more significant than genetic factors, for a ratio of 20:80 and perhaps even 10:90. In 2009, Nisbett, reflecting a more representative distribution

---

[29] Until recently, most professionals believed that the intelligent behavior of individuals is largely determined by genetic factors (Snyderman and Rothman, 1988). This once widely-held opinion may well be changing.

of SES, concluded the ratio to be 40:60. This is to say, as one researcher explained, "performance on IQ tests is most 'heritable' in children of high socio-economic status, perhaps because they have opportunities and support networks to develop their innate talents [more] fully—but that disadvantageous environments prevent children from reaching [as close] their potential"[30]. The same line of research has been characterized as demonstrating that "the underrepresentation of low-SES individuals in behavioral genetic studies" biases those studies "by omitting the portion of the distribution for which environmental effects are known to be the strongest" (This is documented by Nisbett and co-authors in 2012 on pp. 134–5, citing the work of Turkheimer, Harden, D'Onofrio, & Gottesman, documented in 2009.) However, acknowledging their assumptions, findings of each camp has credibility[31].

While these studies address the question of determining genetic and environmental influence on intelligent behavior, and they claim definitive results, in this author's opinion, they do not directly resolve the ratio of environmental and genetic influence on intelligence. The findings of all groups of research are confounded by different levels of environmental influence. At the core of this ongoing discussion is the ambiguity mentioned earlier shown graphically in Figure 2. We do not know how to disambiguate the mix of genetic and environmental influences when measuring behavior. The good news, however, is that, collectively, these researchers unequivocally agree on a basic message namely, *environmental constraints have a substantial effect on intelligent behavior* and, we would add that the environment has far more influence on intelligent behavior than previously realized.

In developing his ratio, Jensen appears to have not given sufficient recognition to the effect of environmental factors including adequate nutrition and good health. As these environmental factors are subsumed, their effect is hardly acknowledged as part of the environmental influence, though they certainly have a strong influence.

---

[30] C. I. Bargmann, personal communication, November 23, 2011.

[31] Precise ratios have not been generated because environments cannot be adequately controlled; twin siblings are not likely to be experimentally separated at conception and carried by surrogate mothers, and, more generally, research is often inhibited, appropriately, by ethical considerations. Thus, definitive ratios of genetic and environmental influence on intelligence are not likely to be available any-time soon.

It is important to expand on this for a moment. Let's assume the validity of several studies concerning the influence of just nutrition and well-being on intelligence which were reported earlier in this book and will be demonstrated later in this chapter. For example, let's assume the validity of (a) providing mineral and vitamin supplements to a disadvantaged pregnant woman increases the IQ of her baby by 8 points, and (b) providing mother's milk to her baby improves the IQ of the baby by another 7 points, together improving her babies IQ by an average of 15 points (8 + 7). (Note that these interventions are independent and thus additive). If increasing a *disadvantaged* child's IQ 15 points by providing mineral and vitamin supplements prenatally, and, by providing mother's milk instead of formula perinatally, then an *advantaged* child's IQ would likely be decreased by about the same amount if the mother were inadequately nourished prenatally and if the baby were provided formula instead of mother's milk.

Stated crisply, **insufficient nutrition is just as detrimental to the disadvantaged as sufficient nutrition is beneficial to the advantaged**.

This notion is so central it demands yet additional clarification. As stated, both genetic and environmental factors determine intelligence. Of that there is no dispute. Furthermore, environmental factors significantly determine the expression of genetic factors. Of this, too, there is no doubt. Thus, crucially, while constrained well-being, typical among the disadvantaged, limits intelligence by constraining, for example, the expression of genetic factors, stronger well-being, typical among the advantaged, supports a stronger expression of genetic factors. Additionally, the more adequate the nutrition the more effective the neuronal system for both the advantaged and disadvantaged. This is likely true for all stages in life. Notably, the test results described previously, using individuals with average socio-economic-status, subsume the environmental benefits that permit cognitive genetic factor(s) to be more extensively expressed and the neuronal system to perform more effectively. Thus, those tests, among the advantaged, do not reveal an appropriate ratio of genetic and environmental influence on intelligence; the benefits are subsumed. In contrast, the tests using individuals with low socio-economic-status do reflect the environmental constraints common among the disadvantaged; genetic influence appears less pronounced. In both cases, environmental factors, likely, influence intelligence *to the same extent – one positively and the other negatively*; **environmental factors, when**

adequately available, influence the advantaged positively to about the same extent as they influence the disadvantaged negatively when inadequately available.

Conveying this a little differently, providing adequate nutrition to disadvantaged individuals raises IQ and then taking away adequate nutrition from the same disadvantaged lowers IQ by the same amount. Why would that be any different for the advantaged? Taking adequate nutrition from advantaged individuals lowers their IQ and then providing adequate nutrition to the same advantaged people raises IQ by the same amount. This conveys that having environmental benefits among the advantaged should have the same influence on IQ, though positively, as not having the same environmental benefits among the disadvantaged, though negatively. Rational reasoning permits one to claim that reducing adequate nutrition from a group of advantaged people would reduce their IQ by essentially the same amount as subsequently providing adequate nutrition would raise their IQ. This claim would not be allowed to be tested for ethical reasons.

More effective expression of genetic factors allows more effective intelligent behavior in the moment and, thus, also more effective accumulation and integration of new information used for subsequent intelligent behavior influencing not only intelligence in the moment but also intelligence growth.

It is therefore concluded that differences in intelligence among humans are largely determined by the environment. Surely, some individuals at the extremes are more influenced by genetic factors but for the clear majority, *differences in intelligence among humans are largely determined by the environment perhaps in the ratio of 20:80, or even lower, as Turkheimer and associates have indicated.*

In net, *environmental constraints such as limited nutrition, restrain genetic material from being fully expressed, and restrain the effective accumulation and integration of information, which inhibits growth of intelligence, and thus forms a limited foundation for subsequent intelligent behavior. Environmental constraints cause IQ to steadily decrease, especially during the developmental stages.*

Therefore, it has become increasingly clear that environmental interventions could enhance intelligence, especially and most substantially of the disadvantaged. Greater possibilities and more opportunities become revealed to those affected, to each one of us, and to society.

These observations are central to the messages of this book. All of us are environmentally inhibited from reaching our potential intelligence, some of us much more than others. We know that reducing those environmental constraints will permit the increase of intelligence of each one of us, and some of us much more than others. This theme is threaded throughout the book.

Additionally, the above reported, though less than definitive, ratios only relate to the intelligent behavior of *individuals*, though they, all too often, are also used, in error, to explain differences in intelligence between *groups of individuals*. This has resulted in yet another myth, one of a set of myths about intelligence, that must be dispelled. This is discussed now, and with other myths, in Chapter 7.

What causes group differences in intelligent behavior?

The possible causes of differences between two identifiable groups of people are those known factors that differentially or disproportionately affect each of the two groups, either by a stronger direct effect on the members of one group, or by affecting disproportionately more members of one group, or both.

While intelligent behavior of any individual human-being, as measured by standard IQ tests, is determined by both genetic heritage and by the individual's past and present environment, these factors permit, promote, or impede the development and expression of the capabilities tested. When the results of IQ tests of individual Americans are grouped and compared by ethnicity and age group, the average IQ of blacks is somewhat lower than whites by age four, and steadily drops from age four to age twenty-four, while the average IQ for white Americans shows little if any change over the same age range. What might explain this data? Which of the factors known to determine individual IQ are possible causes of this pattern of differences?

Complex interrelated environmental factors are known to affect intelligent behavior. These include physical as well as emotional well-being, influenced in turn by health (for example, prenatal and post-natal care, and tobacco, drug use, and alcohol use during pregnancy) and a wide array of nutrients (for example, minerals, vitamins, and proteins). Additionally, formal and informal learning experiences affect intelligent behavior. Other factors include motivation, expectations, and encouragement. Cultural orientations also influence intelligent behavior. Many of the poor experience hopelessness which limits the effectiveness of motivation, expectations, and encouragement.

When children receive a steady adequate diet, perhaps including nutritional supplements, learning experiences are better attended to and absorbed, permitting greater age-appropriate growth of intelligent behavior. When children are poorly nourished or ill, intelligent behavior in the moment and the ability to benefit from formal and informal learning experiences are inhibited. Every day these constraints persist, they impair more opportunities for age-appropriate growth of intelligent behavior, with the cumulative effect that the IQ of children falls, gradually and steadily, further and further behind unimpaired peers. This explains the gradual and steady reduction of IQs of blacks and the disadvantaged during their developmental stages.

In May of 2015, Kimberly G. Noble and associates published an article in the journal Nature Neuroscience entitled *Family income, parental education and brain structure in children and adolescents.* They wrote,

> Here, in the largest study to date to characterize association between socioeconomic factors and children's brain structure, we found that parental education and family income accounts for individual variations in independent characteristics of brain structural development in regions that are critical for the development of language, executive functions and memory. ... [They continue] that parental education was linearly associated with the children's total brain surface area [whereas] ... Family income was logarithmically associated with surface area, implying that for every dollar in increased income, the increase in children's brain surface area was proportionately greater at the lower end of the family income spectrum.

Stated differently, improvements in SES at the low end of the SES spectrum result in much larger improvements in brain structure and surface area than the same size improvements in SES at the high end of the SES spectrum. Also, that a year more parental education resulted essentially in the same improvement in their children's brain surface area across the SES spectrum. While brain surface area is significantly affected by parent's education, this suggests that brain surface area of a child in a disadvantaged family is most affected by family SES. This is distinct from parenting education.

Furthermore, they wrote

> it is unclear what is driving the link between SES and brain structure ... If this correlational evidence reflects a possible underlying causal relationship, then policies targeting families at the low end of the income distribution may be most likely to lead to observable differences in children's brain and cognitive development.

In strong support of these findings, this book reports many studies that explain *causes* of this correlational data. The identified and quantified primary causes include nutrition, health, effective formal and informal learning experiences starting as early as an infant and continuing through college-age.

The comments by Ratey, indicated shortly, report that brain growth and neurological development reflect nutritional well-being and also formal and informal learning experiences. For example, the more a person devotes to playing or practicing the violin, the more area in her cortex is devoted to controlling the activity.

Lynn's comments, indicated shortly, highlight the larger observation: brain size is determined by experiences and well-being and significantly affect intelligence.

So, by examining the factors that affect intelligent behavior at different points in individual human development, we can better understand how group differences in average IQ emerge. For IQ scores to remain constant, as an individual ages, intelligent behavior must grow age-appropriately. The growth of intelligent behavior depends on learning experiences and on the adequate nutrition and well-being that make learning experiences appropriately effective. The ratios often used to characterize the relative importance of genetic and environmental influences on IQ are differently valid for individuals whose environment provides adequate resources for the development of intelligent behavior than for individuals whose environment provides inadequate resources of the development of intelligent behavior. When two populations are compared, and one group experiences more economic disadvantage than the other, differences in average IQ (and also intelligent behavior) are likely to significantly result from environmental factors. Let's now turn to discuss what part of differences in intelligence between two groups that may be caused by genetic factors.

While much has been written about human intelligence that is valid and useful, much has been written that has misled, misinformed, and resulted in great injustices affecting tens of millions in the United States and billions around the world. There is no question that intelligence of each one of us is partly determined by genetic factors and partly determined by environmental factors. Professionals do not dispute this observation. One dispute has been, and remains, on the extent by which genetic factors and environmental factors influence differences in intelligent behavior between *groups of individuals*. While there is abundant evidence that differences in intelligent behavior between **individuals** is partly influenced by genetic factors, *there is no evidence—none whatsoever—that differences in intelligent behavior between **groups of individuals** is even partly determined by genetic factors*. However, while no evidence supports the notion that genetic differences determine differences in intelligence between groups of individuals, there is no evidence for the opposite, that there are no genetic differences between groups which affect intelligent behavior. Of course, in time this may change; evidence of genetic differences between groups of people may surface. But until that happens, it seems reasonable and rational to presume that genetic factors do not determine differences between groups of people.

So, we may now convey explicit comments regarding the factors that influence differences in intelligence between groups of individuals. There is overwhelming evidence that supports the awareness that differences in intelligence between groups of individuals such as the advantaged and disadvantaged, or, blacks and whites, are determined by environmental factors such as nutrition, health, and the resulting effectiveness of formal and informal learning experiences. This understanding should change long-held notions that importantly have affected several profound and far-reaching societal problems; injustices perpetrated by long-held and false notions that genetic factors play a role in group differences of intelligent behavior continue to this day. Other myths of intelligence are conveyed in Chapter 7.

Reducing environmental constraints, especially of the disadvantaged, enhances intelligence

With that background, this book conveys an understanding of environmental constraints, their impact on intelligent behavior, as well as their effect on societal problems.

- We review research on nutrition and health factors that limit the well-being of the neuronal system and are largely related to economic disadvantage. This is the focus of this chapter.
- We project the qualified effects of those factors on the IQ deficit associated with economic disadvantage, the focus of this chapter, as well as the quantified effects on the advantaged and disadvantaged IQ gap, as well as the black–white IQ gap, the focus of Chapter 5.
- We reflect on how improvement of those factors might alter the IQ Bell Curve and substantially reduce the societal problems such as school dropout rate and employability associated with individuals with relatively low IQ scores. Chapter 6 presents this discussion.

Can intelligent behavior, especially of the disadvantaged, be enhanced? It certainly can, and considerably! The evidence is clear and not in dispute. Intelligent behavior may be enhanced, most significantly of the disadvantaged, by reducing nutritional and other environmental constraints on individuals so that each may more closely reach his or her potential intelligence. By way of emphasis, enhancing one's environment does not, in and of itself, improve intelligence. For example, improving the nutrition of adequately nourished people hardly changes their intelligence. Rather, environmental constraints, for example, being undernourished, limits the development of intelligence and thus improving one's environment, especially of the disadvantaged, reduces the constraints on the development of intelligence. Not so remarkably, intelligence of most of those in the lowest 20 percent of the spectrum can be raised into, at least, the near-average range[32].

This book shows how, and to what extent, enhancing intelligent behavior, most significantly among the disadvantaged, reduces the opportunity gap, improves upward mobility, permits people to more closely achieve their potential, permits them to become more productive members of society, and to achieve human dignity. Just raising intelligent behavior, at least of the disadvantaged, increases their capability, the collective human capability, and therefore substantially reshapes and redefines societal problems.

The focus, here, is to reduce inequality of opportunity, which will, consequently, reduce economic inequalities. The intent is to identify the constraints on opportunity most prevalent among the disadvantaged. It is shown that economic inequalities[33] will be reduced as constraints on opportunity are reduced. Of course, constraints on opportunity take many forms. We can now move on to discuss ways to influence intelligence.

Assuring adequate nutrition, prenatally and in early childhood, enhances intelligence

The measures of intelligent behavior currently in use for children younger than about four-years-old are not directly comparable to those used for older children. In keeping with this pattern, the effects of prenatal, perinatal, infant diet, and well-being are evaluated here in terms of their effects on subsequent intelligent behavior at and after age four.

In 1998, Martorell suggested several explanations as to why the IQ gap between disadvantaged and advantaged children widens to the age of four. He writes,

> Nutritional problems are most common and severe in the
> first 2-3 years of life for several reasons (Martorell, 1995).

---

[32] This phrase, *near-average range*, is less than well defined, but in school systems the phrase often suggests an IQ between 90 and 110.

[33] A separable problem may be mentioned. Why have economic inequalities become increasingly worse in recent decades? Many reasons may be offered. And though they certainly have aggravated the problem and need close examination, these most recent exacerbating reasons are viewed as consequences of economic inequalities. Pursuing them may be a digression here, and thus are postponed to later in the book.

Young children have greater (i.e., per unit of mass) nutritional needs, in part because they are growing faster than at another point in life, including adolescence. At the same time, young children have relatively small gastric capacities and thus require frequent meals with rich nutrient concentrations; however, in many poor settings, children receive infrequent dilute or bulky diets of low nutritional value. ... Another reason for the vulnerability of small children is their complete dependence on others for care. These factors, along with intrauterine growth retardation, account for the fact that many nutritional problems are at their peak of incidence and severity in early childhood. (pp. 185–6)

The growing brain particularly demands nutrients, as Benton reminds us in 2011,

At birth although the infant is about 6% of the adult body weight, the brain is already 25% of its final weight. By 2 years of age the brain will be about 77% of its final weight while the body is about 20% of the adult level [(Dekaban & Sadowsky, 1978)]. Such a rapid rate of growth inevitably places demands on the diet. (p. 15)

These comments by Martorell and Benton emphasize the importance of diet in supporting the growth of the brain during the earliest developmental stages.

Also, brain growth and neurological development are known to not only reflect nutritional well-being but also formal and informal learning experiences. In 2001, Ratey explains:

An accurate map of the brain would be different for each of us and would shift over time. Connections that receive input from frequently used body parts will expand and take up more area [in the brain] than those that receive input from infrequently used parts. Magnetic resonance imaging (MRI) shows that the brains of violin players devote much more area to pathways representing the thumb and fifth finger of the left hand—the fingering digits—which are used extensively in hours of training. The younger a child begins practicing, the more area her cortex devotes to these fingers (p. 35).

These factors play a role in affecting the average intelligent behavior of whole populations. In 1998 Lynn explains:

the crucial point... is that [along with IQ] head size and brain size have... increased during the last half century (Miller and Corsellis, 1977; Ounsted, Moar, and Scott, 1985). The significance of this fact is that brain size is a determinant of intelligence. What has occurred is that the improvements in nutrition have increased the growth of the brain, and probably also its neurological development, and this has increased intelligence. (p. 211)

I hasten to add, while brain size influences intelligence, women's brains, on average, are smaller than that of men's yet their respective intelligence, as measured by IQ tests, is essentially the same. Apparently, there are additional factors that determine intelligence. The relevant observation is that with improved nutrition women's and men's brains are larger affecting their respective intelligence.

More recently, in 2015, Mackey and associates documented research that builds on these observations. They wrote

In the United States, *the income-achievement gap* –the difference in academic achievement between students from higher- and lower-income backgrounds – is substantial and growing (Reardon, 2011). [With that as background, Mackey and associates reported that] cortical thickness in all lobes of the brain was greater in students from higher-income than lower-income backgrounds. [And that] Greater cortical thickness, particularly in temporal and occipital lobes, was associated with better test performance. (p. 1)

They continue,

> Neither the causes nor the cellular bases of differences in cortical thickness are known. [However,] low SES is associated with many factors that influence brain development, including enhanced exposure to stress and reduced environmental enrichment (Hackman & Farah, 2009). ... Critically, neuroanatomy is modifiable through experiences. Neuroimaging studies have shown changes in brain structure after a few weeks of learning (Zatorre, Fields, & Johansen-Berg, 2012). Therefore, educational programs may positively influence neuroanatomical circuits that support cognitive abilities. For example, an intervention that involved both children and parents was shown to enhance electrophysiological brain measures and cognitive functions in younger children from lower-income backgrounds (Neville et al, 2013). (p. 8).

Importantly, they write,

> The lower-income group had a larger proportion of racial and ethnic minorities, as characterizes lower-income groups in the United States, but *neither race nor ethnicity explained significant variance in cortical thickness in the regions that differed significantly between income groups.* (p. 6).

Italic emphasis is mine.

Stated differently, larger cortical thickness is not due to racial or ethnic differences but to differences in economic background.

Several conditions frequently noticed among the disadvantaged and experienced prior to age four relate to lower subsequently measured IQ: poor maternal nutrition during pregnancy, short length of gestation, low weight at birth, not being breastfed or being breastfed for too short a period, and short interval between the births of siblings. Higher proportions of black children than white children are born and raised in conditions of poverty or relative poverty and are therefore disproportionately exposed to these factors that inhibit intelligent behavior in the moment and the growth of intelligent behavior over time.

Worth noting, yet again, except where highlighted, comments in this chapter do not relate to racial differences but rather relate to differences in environmental constraints experienced by the advantaged and disadvantaged.

Assuring adequate nutrition of pregnant women and, also, prenatal nutrition of their babies

A pregnant mother's nutritional well-being significantly affects not only her baby's health but her baby's subsequent intelligent behavior as well. For example, as reported in 1955 by Harrell and associates, giving vitamin and mineral supplements to disadvantaged pregnant women results in their offspring having, on average, an eight-point higher IQ (p. 64). It is noteworthy that pregnant women, whether economically advantaged or disadvantaged, who receive prenatal medical care are currently advised to use such nutritional supplements. This can be achieved at a cost of pennies a day. So, it is recommended that mineral and vitamin supplements be made accessible to all pregnant women, ideally in the context of regular medical and dental care.

The definition of the supplements depends on various factors that may be somewhat generalized for various disadvantaged groups. For example, Lynn reported in 2006 that since "the effect of severe iodine deficiency is to reduce intelligence by 13.5 IQ points" (p. 183), groups in central Africa might need more iodine than groups living on the coast of Africa where seafood, abundant in iodine, is more plentiful. Other variations exist in the United States as well as different parts of the planet.

Mineral and vitamin supplements merely enhance the well-being of pregnant women and indirectly their offspring toward the average. They do not provide enough enhancements to reach average well-being because proteins and other nutrients are not included in these supplements. (The postnatal enhancement studies to be discussed later assume average prenatal well-being.)

Increasing the likelihood that babies be breastfed for an optimal period

Extensive evidence, now widely acknowledged, supports the claim that breastfeeding significantly affects the health and well-being of children.

Breastfeeding also significantly affects intelligent behavior. Providing babies with mother's milk rather than formula can increase their subsequently measured IQ by 3.3 to 8.2 points for females and by 1.0 to 5.8 points for males; the lower increases are associated with breastfeeding for three weeks or less; the larger improvements with breastfeeding for 6 months. (This according to Quinn et al., as reported in 2001; cf. Nisbett et al., 2012, p. 136). There is reason to believe this increase results from nutrition, not from nurturing touch.[34] So it is recommended that, through parenting education and other means, mothers be encouraged to breast-feed children up to six months of age.

Another important study, a meta study encompassing 17 studies, reported in 2015 by Horta and Associates, conveys that all 17 studies support the notion that providing mother's milk to infants improves their IQ. Their reported average increase is 3.44 IQ points. Because this average reflects several confounds, we chose to use Quinn's reported average of 7 points; Quinn's article is more focused on the relevant factors. For example, the Horta meta-study includes studies in which children's IQ were tested between age 2 and 19, whereas the Quinn study focuses on 5-year-olds. This is significant because, as indicated by Horta and Associates

> [s]tudies that evaluated subjects aged between 10 and 19 years ... reported a smaller benefit from breastfeeding [mean difference: 1.92 points ...] than studies involving younger subjects [mean difference: 4.12 points...].

Much of these differences may be explained by the decrease of 0.6 IQ points every year among disadvantaged children, regardless of race, from age 4, as shown previously. Also, the Horta study includes a wide difference in mother's milk provided infants from *ever* breastfed to more than 8 months breastfed, whereas the Quinn study focuses on children breastfed for 6 months or more.

---

[34] "Fatty acids are essential for brain growth and efficient functioning; about half of these acids are acquired in utero and the other half in the first 12 months of life from breast milk; these fatty acids are not present in cow's milk or in most infant formulas, which is one reason why infants who are breast-fed have higher subsequent IQs" (This is documented by Lynn, in 2006, p. 187, citing Grantham-McGregor, Walker, and Powell, 1994).

According to Victora and associates[35], 832,000 children, worldwide, die before the age of five every year because they were not fed mother's milk as infants. As huge as this appears, it is only a portion of the tip of the iceberg. This enormous tragedy is only a trivial part of the tens of millions of youth who live their life in poor health and limited capability because they were not fed mother's milk in the first half year of their life. Yet earlier this year US delegates at the World Health Organization, WHO, tried to prevent fostering mother's milk over formula perhaps to protect the all-important profits, more than $40 billion, of formula suppliers.

According to Jacobson and Jacobson, writing in 2006, nurturing during breast-feeding enhances well-being and intelligent behavior. Yet, revealingly, giving mother's milk to low-birth-weight (LBW) babies increases their subsequent IQ by an average of 10.2 points, as reported by Johnson in 2001, *even when the babies are fed through a tube—without nurturing.*

Low weight at birth inhibits intelligent behavior

Studies show that LBW babies are about 13 points lower in IQ, on average, than are normal birth weight (NBW) babies. This was reported by Sandra Scarr, in 1981, p. 52. Ortiz-Mantilla and associates reported in 2008 that very low birth weight (VLBW) babies

> performed more poorly on visual and auditory-visual habituation tasks and scored lower than controls on cognitive and language measures. [These] differences in language abilities ... may be part of a global deficit that impacts many areas of cognitive functioning rather than a specific impairment in rapid auditory processing.

Additionally, it is known that VLBW babies have far more serious medical conditions and are accordingly likely to show significantly lower intelligent behavior than LBW babies— that average intelligent behavior of not-so-LBW babies is nearer that of NBW babies, but likely to be below average in intelligent behavior. Because there is no specific data showing the likely

---

[35] Victora and Associates (2015)

average IQ for VLBW or the not-so-LBW babies, these values are estimated. This book assumes the IQ deficit to be ten points more for VLBW than for LBW babies (twenty-three points instead of thirteen) and ten points less for not-so-LBW babies (three points instead of thirteen). These observations are considerably expanded and quantified in the next chapter.

In 1998, Jensen wrote that providing nutritional supplements to pregnant women is likely to reduce the frequency of LBW babies. Reducing teenage pregnancy rates would also reduce the frequency of LBW babies: women who are still growing absorb nourishment for themselves to the detriment of their babies. So, it is recommended that, through parenting education and other means, teenage girls be appropriately informed of these observations and encouraged to delay pregnancy, and that all pregnant women be encouraged to maintain an adequate diet.

It should be noted that a distinction is made between premature babies, who may not be LBW babies, and babies who are born weighing less than 2,500 grams. Also, it is important to recognize that there are many different reasons for babies to be born with low birth-weight or prematurely. We do not consider all these different possibilities.

The costs of intervention are entirely offset by immediate savings in related areas

Let's digress for a moment. Though this was mentioned earlier we feel more detail clarifies the issue of offsetting costs of interventions.

In the United States 330,000 LBW babies are born each year. The disadvantaged have twice as many LBW babies as the advantaged; about 220,000 LBW babies are born to disadvantaged women each year while advantaged women have 110,000 LBW babies. Caring for *each* of these LBW babies until they are released from the hospital costs society more than $50,000. The average cost for disadvantaged LBW babies is even higher because the disadvantaged give birth to these babies significantly earlier and substantially lighter in weight than do the advantaged. Thus, society pays *more than* about $11 billion, in only the United States, for just the initial hospitalization care for the LBW babies of the disadvantaged. Let's look at that more closely.

First, let's assume that the causes of the 110,000 LBW babies born to the advantaged have essentially the same causes as half, 110,000, of the LBW born to the disadvantaged. That's not likely valid, but that assumption permits us to make some useful conclusions. Now let's focus on the other half of the LBW babies born to the disadvantaged. What would explain these additional 110,000 LBW baby births?

As will be seen in Chapter 6 Figure 27, evidence shows that the women who have 45 percent of LBW babies have IQs in the lowest quintile of the IQ spectrum; women who have 80 percent of LBW babies have IQs in the lowest half of the IQ spectrum. Why would intelligence, as measured by IQ tests, explain the frequency of LBW baby births? Is the explanation perhaps really the environmental constraints on intelligence including inadequate nutrition and poor health typical of those with limited intelligence? In fact, the National Center for Health Statistics indicate

> Mothers in deprived socio-economic conditions frequently have low birthweight infants. In those settings, the infant's low birthweight stems primarily from the mother's poor nutrition and health over a long period of time, including during pregnancy, the high prevalence of specific and non-specific infections, or from pregnancy complications underpinned by poverty. Physically demanding work during pregnancy also contributes to poor fetal growth[36].

A. Providing mineral and vitamin supplements to disadvantaged pregnant women not only improves the IQ of their offspring, as mentioned, but also, because each of the disadvantaged pregnant women is better nourished, reduces the likelihood of the offspring being an LBW baby. The cost to provide mineral and vitamin supplements to all disadvantaged pregnant women would be about $15 million[37] per year. That's 2 pennies a day[38] for each disadvantaged pregnant woman. The cost of this intervention is far more than

---

[36] NA NCHS (National Center for Health Statistics), National Vital Statistics report 2002, Vol. 51.

[37] 4 million babies born in the US per year, some of which are multiple births e.g., twins. ½ of their mothers live in poverty or near-poverty. Cost is $10 for mineral and vitamin supplements per year. Actually, 9 months or ¾ year.
4 million (½) $10 (¾) = $15M per year,

[38] Cost per day per disadvantaged woman is overall cost divided by number of babies born to disadvantaged women divided by days of the year or $15M/2M/365 = $0.02

offset by the savings resulting from just the reduction in LBW baby births. This $15 million represents 0.27 percent, or the medical costs of about 1 in 400 LBW babies.

Furthermore, this intervention would also reduce the severity of subsequent problems for each LBW baby. Providing mineral and vitamin supplements to disadvantaged pregnant women also significantly improves the general health of both the mother and her baby thus reducing other subsequent health care costs. This intervention raises the IQ of the offspring whether an LBW or NBW baby. Incidentally, as mentioned, advantaged pregnant women receiving prenatal medical care are prescribed mineral and vitamin supplements for health reasons.

B.   In contrast with advantaged school children whose IQ remains steady throughout their school years, the IQ of disadvantaged school children drops at a steady rate of about 0.6 points per year or 12 points from preschool through age 24. Providing mineral and vitamin supplements to all disadvantaged children, age 4 to 24, would substantially prevent their IQ from steadily falling these years. The evidence for this is provided by Herrnstein and Murray. The annual cost of this intervention would be about $544 million[39] (about $4.46 per US tax payer[40]). (As another data point, this $4.46 per taxpayer compares to $12,509 spent per year on each public-school student[41].)

If this steady diet of mineral and vitamin supplements throughout the school years reduces the school dropout rate by only one percent and therefore the under-employability rate and likely criminality by only one percent, it would prove to be a great bargain for society. As will be seen in Chapter 6, the likely reduction in school

---

[39] (US population) x (% of those in poverty or near poverty) x (% of population between age 3 and 18) about $10 per year for mineral and vitamin supplements or 330M(.5)(.33)($10) = $544M/year

[40] $544M/number US taxpayers or $544M/122M = $4.46 per US taxpayer

[41] U.S. Department of Education, National Center for Education Statistics. (2017). The Condition of Education 2017 (NCES 2017-144), Public School Expenditures.

dropout rate, under-employability, and criminality would be much higher.

As to the cost to society for just these two interventions, if observation (A), mentioned just above, were to decrease the frequency of LBW babies with a commensurate increase in the frequency of NBW baby births by one in twenty, then these collective interventions would cost society nothing; the cost-savings resulting from reduction in births of LBW babies would pay for providing mineral and vitamin supplements daily to all disadvantaged children from age four through age twenty-four, thus significantly increasing the IQ of disadvantaged children, nearly enough to eliminate the IQ gap between the disadvantaged and advantaged.

A simple study could provide evidence to support this hypothesis. Namely, supply disadvantaged pregnant women with mineral and vitamin supplements every day of their pregnancy – perhaps as early as when they decide to seek to be pregnant. If the frequency of LBW baby births is decreased by one in twenty, then the observation, the hypothesis, is supported.

The reduction in major medical problems and their cost offset the cost of interventions that reduce the environmental constraints which, in turn, increase intelligence, especially of the poor.

Multiple births and births in close succession constrain early cognitive development

As Jensen conveyed in 1969, p. 66-67, identical twins (MZ) and fraternal twins (DZ) are known to have lower IQs, on average, than single-births. Twins, for example, are known, on average, to have IQs of 6 points lower than that of single-births. The lighter twin has 5.5 points lower IQ than the heavier one. In 1999, Bodmer (p. 335) added, that on average, triplets have four points lower IQ than twins. Multiple birth children share resources both in their prenatal environments as well as their post-natal environments.

Births in quick succession, about a year apart, Hunt reported in 1961, also have slight but a somewhat more significant effect on average intelligent behavior. A similar reduction in IQ is shown for births in close succession, about one year apart, as for multiple births. This is not true for people who have help with child care. The help, perhaps family, seems to allow the children to receive adequate resources, attention, and affection after birth so they thrive more similarly as do single-births.

Hunt also, in 1961, explains that

> inasmuch as about the same degree of inferiority appears for twins and for doubles born close together as appears for those in large families, and inasmuch as this negative correlation disappears among the well-to-do who can afford to have help in the care of their infants during the early months, it [appears that strong intelligent behavior fails] to develop from lack of the stimulation that comes from young infants having ample adult contact. (p. 362)

Research of Sulloway, in 2007, show birth order also affects IQ scores, with the IQ of the first born in a family tending toward four points higher than the last born.

Parental stimulation, nurturing, caring, expectations, and simple attention appear to affect intelligent behavior. So, it is recommended that, through parenting education and other means, parents be made aware of how births in close succession and multiple births affect well-being and intelligent behavior, be encouraged to allow more time between births, and to find sufficient time for each child. It is worth emphasizing that these parental activities take time, and while these can be achieved with babies in close succession, the demand on the parents can be difficult.

So far, we have only considered a subset of influences on intelligent behavior. Much of prenatal and perinatal care has not been addressed, including, for example, fetal alcohol syndrome (FAS), smoking, and prenatal and perinatal medical care. These are not addressed here, despite their relevance, partly because adequate information is often not available to quantify their effect on differences in intelligent behavior between the economically disadvantaged and advantaged. This is discussed more fully later in the book.

Other conditions required for the growth of intelligent behavior, up to age four, will be discussed when we take up the question of adequate nutrition during formal and informal learning experiences.

The prenatal and perinatal effects on intelligent behavior just discussed are, for the most part, independent and cumulative, as the studies cited explicitly establish. Providing nutritional supplements to economically disadvantaged pregnant women improved their offspring's well-being, but this improvement was limited by the fact that only mineral and vitamin supplements were supplied; no protein or other nutrients supplements were provided. Therefore, while the child's well-being was likely increased by not quite enough to reach the average, the studies on the effects of providing mother's milk to babies assume average well-being at birth. The nutritional supplements provided prenatally only help the babies nearly reach average well-being. Thus, the effect on intelligent behavior resulting from providing prenatal mineral and vitamin supplements may be viewed as entirely cumulative with the effect resulting from providing mother's milk to babies.

Considering another separable aspect of this analysis, LBW deficits have a strong effect on the *total* population IQ averages. Also, the effects of the various types of LBW births somewhat overlap with effects of nutritional deficit or nutritional supplements as well as with the effects of breast-feeding babies.

Providing economically disadvantaged pregnant women with mineral and vitamin supplements would, likely, reduce the frequency of LBW babies, as indicated by Jensen in 1998. Also, it is known that providing mother's milk to LBW babies increases the babies' average IQ by 10.2 points thus reducing their overall deficit from 13 IQ points to about 3 points. This potential intervention would likely reduce the effect on intelligent behavior of being a LBW baby.

This discussion demonstrates the independence of effect among two factors, maternal supplemental nutrition provided disadvantaged pregnant women, regardless of race, and providing more mother's milk to babies, again, regardless of race. It also demonstrates that the effects of low birthweight somewhat overlap with the other two factors.

Note that other considerations, such as formal and informal learning experiences by age four, were not considered so far in this book because of limited direct data. Some data that would support estimating the effects of those factors are included in the next section. The relevant studies do not focus on, but sometimes do not exclude, the years to age four. Thus, the effects of learning experiences are not itemized for these earliest developmental years, and their effects during those years may be included in the effects discussed for age five and older.

Assuring adequate nutrition and effective learning experiences from age four to eighteen

Intelligent behavior may be viewed as determined by three distinct sets of factors: 1) cognitive and memory capabilities (the *potential* for which is significantly determined at conception), 2) maintenance of these capabilities via adequate nutrition, health, emotional well-being, and, 3) the growth of information accumulated via formal and informal learning experiences and gradually amassed in the memory structure(s). The previous section examined the maintenance of cognitive capability through early childhood. This section examines factors affecting the accumulation of knowledge through early adulthood.

Formal and informal learning experiences contribute to the growth of intelligent behavior. These experiences are less effective when individuals are less than adequately nourished. Year after year of relatively poor diets among economically disadvantaged children and adolescents add up to a cumulative lag in information acquisition; that is, a cumulative deficit in growth of intelligent behavior.

Attending school undernourished inhibits the benefits of attending school; those attending school with adequate well-being can be expected to gain more from learning experiences; those with inadequate well-being will likely increasingly fall behind. Age-appropriate growth of intelligent behavior depends on learning experiences under conditions of adequate nutrition and health—adequate well-being.

In one "socially deprived city in Britain" studied recently, 17 percent of adolescents were iron deficient. Lynn reports[42] in 2006 (p. 182) that iron deficiency reduces

> the number of dopamine receptors and this impairs dopamine neurotransmission, which in turn impairs learning

---

[42] Citing Lynn and Harland, 1998; cf. Lynn, 2006, p. 279

and brain function in adulthood....[D]aily iron supplements given... for three months increased [the teenagers'] IQs by 5.8 IQ points.

Eysenck, reported in 1991 that nutritional deficiencies, specifically deficiencies of certain vitamins and minerals essential for optimal brain development and cognitive function, cause a reduction in IQ of about eleven points.[43] Meta-analysis of nutritional supplements provided to individuals with vitamin deficiencies, by Benton in 2001, concluded that supplementation appeared to improve performance on tests of nonverbal measures of intelligence but did not improve performance on verbal measures. Benton in 2011 argued,

Such a highly selective response suggested a genuine phenomenon, particularly as such a response is theoretically predicted. There is a distinction between crystallized and fluid intelligence [(Horn & Cattell, 1966)]. Fluid intelligence is the ability to reason; an ability that is independent of education and experience that is perceived to reflect basic biological functioning. In contrast, crystallized intelligence reflects past experience and involves verbal ability and acquired information. Fluid intelligence can be assessed using non-verbal intelligence tests whereas crystallized intelligence is measured using verbal tests. Logically,

---

[43] Interestingly, Jensen, in 1998, adds (p. 507) that this is not unique to the economically disadvantaged. Evidence shows that dietary deficiencies, mainly in certain minerals and trace elements, occur not only in economically disadvantaged groups but even in some middle-class white families that enjoy a normally wholesome diet and show no signs of malnutrition. Blood samples were taken from children in such families prior to receiving supplements of certain minerals to the diet. Only those children who showed a significant gain of nine IQ points, after receiving the supplements for several months, showed previous deficiencies of one or more of the minerals in their blood. This is a clear indication, yet again, that the issue is *not racial*, not exclusively related to the economically disadvantaged though more likely realized among them, and most likely realized among blacks because they are disproportionately economically disadvantaged. Benton (2001, p. 297) concurs, "The evidence is that not all children respond to supplementation, rather there is a minority who benefit, whose diet offers low amounts of micro-nutrients."

given that a micronutrient supplement cannot be expected to increase vocabulary or basic knowledge, whereas it could enhance basic biological functioning, the pattern of findings is exactly as would be predicted. (p. 20).

These increases in non-verbal IQ scores did not apply to those parts of the IQ tests that reflect growth in knowledge (which can only be recognized in tests repeated over months or years). The increases related only to fluid intelligence; that is, to intelligent behavior in the moment. However, intelligent behavior *in the moment* determines the effectiveness of absorbing, integrating, and accumulating the information that will subsequently demonstrate *growth* of intelligent behavior.

So, providing vitamin and mineral supplements to economically disadvantaged children improve intelligent behavior in the moment, and if the nutritional supplements are given over the years, or as a steady diet, the individual is more able to benefit from formal and informal learning experiences, and his or her IQ is less likely to lag.

In this discussion of the K-12 years, we examine the effects of parenting as well as of formal and informal learning experiences. We assume underlying adequate nutrition, so that intelligent behavior may develop and grow over time. For children who are less than well nourished, supplemental nutrition, by itself, directly enhances intelligent behavior in the moment and helps maintain health and strength so that new information is more likely absorbed and integrated. This both maintains well-being in the moment, making previously accumulated information more readily accessible, and permits the growth of intelligent behavior. For adequately nourished children, supplemental nutrition has limited effects on intelligent behavior.

While the importance of nourishment to well-being is widely understood, and while the importance of nourishment to intelligent behavior is generally known by researchers, it is hardly understood by the general-public. And adequate nourishment is only one of the environmental considerations required to enhance intelligent behavior.

Infant training can have dramatic impact on intelligence as measured by IQ tests, even for Down syndrome children. In 1981 Scarr indicates

> To illustrate the phenotypic changes that can be produced by radically different environments for children with clear genetic anomalies, Rynders (Personal Communications, November 1971) has provided daily intensive tutoring for Down's syndrome infants. At the age of two, these children have average IQ's of 85 while control-group children, who are enrolled in a variety of other programs, average 68. Untreated children have even lower average IQ scores. (p. 72)

Note, this represents an increase in IQ, of Down Syndrome children, of more than seventeen points just for tutoring within only the first two years of life.

Very Early Experiences—environmental influences

While studies mentioned earlier focus on nutritional and health considerations, several studies convey the importance of early overall environmental conditions on intelligence, including formal and informal learning experiences. These focus, though not exclusively, on the first two years after birth.

One out of six births among blacks is an LBW baby. Also, one of twelve births, among whites, is an LBW baby. Disproportionately, many of the white LBW babies are born to disadvantaged women. The same is true for blacks but much more pronounced because blacks are three times as likely to be disadvantaged. This data, alone, significantly affects the IQ of disadvantaged children, regardless of race. It explains much of the IQ gap between the disadvantaged and advantaged, and more of the IQ gap between blacks and whites.

Experiences in early childhood have profound effects on well-being and intelligence. In 1961, Hunt (p. 33) comments that the studies of Spitz in the mid-1940s have had great influence in convincing many people, especially from the professions of psychiatry and social casework, that intelligence is not fixed but plastic and modifiable and that mothering is crucial during the first year after birth. These studies are still referenced by researchers as recently as 2015 including by Nelson of Harvard University, mentioned just below. But first we expand on Spitz's ground-breaking work.

Spitz compared two groups of infants, one in an orphanage in which the children received little attention or variation of stimulation of any kind after having been nursed by their mothers for three months. Hunt says, in 1961, that these "socially well-adjusted mothers" (p. 33) had only one handicap, which was an inability to support their babies. The other group, as indicated by Hunt, consisted of children in a nursery attached to a penal institution (reformatory), of mothers who were mostly "delinquent minors as a result of social maladjustment or feeblemindedness, or... physically defective, psychopathic, or criminal." Hunt (pp. 33-34) writes that the development quotient (DQ) for the sixty-one children in the orphanage dropped during the first year of the infant's life from a starting level of 131 for months 2-3 to a final level of 72 for months 10-12. In contrast the DQ for the 69 children in the nursery attached to the penal institution was about 97 for months 2-3, rose to 112 at months 4-5, remained level to months 8-9, then dropped to 100 for months 10-12. While 37 percent of the children in the orphanage died within two years (two from disease) and more in the following year, all children in the reformatory nursery were alive after five years.

> Spitz attributed the decrease in DQ at the orphanage to lack of mothering. ... Spitz reports that segregation in the penal institution frustrated the mothers' usual outlets, so they lavished their tenderness and pride on their infants each day.

Another orphanage study, writes Hunt in 1961 (p. 361), is also revealing.

Inasmuch as Dennis has very recently found that in orphanage environments where the variety of stimulation is minimal, only 42 percent of the children sit alone at two years of age, and only 15 percent walk alone at four years of age, it appears to be quite clear that the rate of development is not predetermined by the genes.

Or, at least, development is significantly affected by environmental factors—nurturing and its related experiences, especially in this case.

Since Spitz's work, about seventy years ago, much research on the vital and basic need for nurturing was documented. Several researchers including Nelson of Harvard University, in 2014, studied parental influence on cognitive behavior during the critical first two years of life. They extended understanding that the development of the brain is severely affected if parental or caregiver nurturing is inadequate.

Their study involved 136 parentless children in Bucharest whom they randomly assigned either to the usual institutional care available in Romania's overcrowded, poor-quality orphanages, or to foster parents recruited specifically for the project and who had received special training in providing loving care. There was also a control group of never-institutionalized children.

A major conclusion of the study was that high-quality, child-centered foster care is better than institutional care. On many measures, including IQ, the less time a child was in institutional care (under the age of two), the greater the recovery, although in no case was it enough to allow children to reach the level of those reared in parental homes. There were also intriguing changes in measures of brain function pointing to improvement for those placed in foster care before age two. Finally, changes were observed in the molecular signs involving telomere size that have been related to survival in old age. These important results fit with increased evidence in neuroscience for brain plasticity in children and adults, however raised.

As reported by the National Scientific Council on the Developing Child in 2012,

> There is extensive evidence that severe neglect in institutional settings is associated with abnormalities in the structure and functioning of the developing brain. ...Chronic neglect can alter the development of biological stress response systems in a way that compromises children's ability to cope with adversity. ... Children [, furthermore,] who have experienced serious deprivation are at risk for abnormal physical development and impairment of the immune system. ... [Additionally, severe neglect in both family and institutional settings is associated with greater risk for emotional, behavioral, and interpersonal relationship difficulties later in life. ... Children who have experienced severe neglect are more likely to have cognitive problems, academic delays, deficits in executive function skills, and difficulties with attention regulation.
>
> Public concern about the problem of child maltreatment is focused disproportionately on the dangers of physical and sexual abuse, while significant neglect receives less attention. ... [Accordingly, d]espite considerable advances in scientific knowledge about the short- and long-term consequences of significant deprivation and the importance of prompt intervention, most child welfare agencies have relatively limited capacity to address the developmental needs of young children who have experienced reportable neglect. ... Measuring the economic benefits of interventions that improve life outcomes for young children who experience significant neglect or chronic under stimulation would provide important data to justify and guide enhanced resource allocation.

In June of 2014, partly in response to these comments, Iris Adler reported, the cost of not acting early is huge in human terms, and also financially.

According to a recent Centers for Disease Control and Prevention study, "the total lifetime cost of child maltreatment is $124 billion each year." And this estimate does not adequately reflect perinatal neglect as discussed here. Adler continues that the head of Harvard's Center on the Developing Child, Dr. Jack Shonkoff, "says we must respond urgently to the scourge of childhood neglect, as we have to other public health threats." *Do we respond to this insensitive and costly mistreatment of people, appropriately. There is reason to believe that we do not.*

While animal studies do not necessarily have a direct bearing on human understanding, they often provide valuable insights. In this regard Thompson and Heron, in 1954, found that dogs, pet-reared for the first eight months of their lives, perform at a considerably higher level of intelligence at eighteen months of age than do dogs cage-reared for their first eight months. This appears to contribute to understanding human behavior.

In 1981 Scarr found that offspring of lower-IQ natural mothers did not achieve as high IQs as the children of higher-IQ mothers, regardless of adoptive family characteristics. Scarr wrote,

> Eleven adopted children whose natural mothers had IQs of less than 70 (mean = 63) had an average IQ of 104. Eight adopted children, whose mothers had IQs of above 105 (mean = 111) have an average IQ of 129. Although the number of cases is very small, the results suggest (1) that there is a considerable reaction range shown by the children's IQ scores, and (2) the genotypic differences between groups of children with retarded natural mothers and those above-average natural mothers were important in determining the rank order of the children's IQ scores. (p. 43)

She then adds, importantly, that even the children with retarded natural mothers scored well above average, which demonstrates the importance of the adoptive home environments in raising the IQ level of the adopted children. In addition to highlighting that even children of women labeled mentally retarded had above average IQs, another important observations may be made, that being, a reason explaining the lower scores of the children whose natural mothers had IQ scores that averaged 63, likely was both their prenatal and perinatal care which probably was considerably less than that provided by the mothers who had IQs that averaged 111. Prenatal and perinatal care have significant impact on intelligence. It likely accounts for much of the 25-point difference (129 – 104).

Children benefit from their caregivers' early parenting education

Many studies indicate that children cared for with "initiative, attention, and continuity" score higher on IQ tests, on average by "approximately 7 to 11 IQ points"[44] than children who receive poorer quality care. The kind of care includes talking to children, listening to them, touching them, and giving them both physical and emotional care.

While the peers of adolescents surely have a strong influence on them, as is reported by Harris in 1998, it is suggested here and elsewhere that parents, too, have a significant influence over the well-being and intelligence of offspring, certainly pertaining to an individual's early life but also through adolescence. For that reason, Bloom wrote in 1964,

> Perhaps in the future we will find the task of being a mother
> or father requires far more training and preparation than
> has been commonly recognized. (p. 229)

Knowing the importance of parenting, it is time society gave this activity serious consideration. Society should consider classes in parenting in the school systems. It is clear such classes would probably have a profound impact on the happiness and well-being of each family, each individual, and importantly, society.

Where the caretaker is involved and given direct instruction regarding ways to encourage the child, gains are maintained and even built upon[45]. For example, economically disadvantaged "parents provide poorer problem-solving strategies for their children than do middle-class parents, and they tend to solve problems for their children rather than assist them in

---

[44] Flynn, J (1980), p. 179, interpreting the findings in Tizard, Cooperman, Joseph, and& Tizard, (1972)
[45] Bronfenbrenner, (1999b)

solving them."[46] With awareness, many practices may be relatively easily corrected. In general, the observation that family involvement is necessary consistently surfaces.

The effects of preschool experiences and parenting training are independent, as shown by several studies. Eckberg points out in 1979 (pp. 109-110) that the largest, and most long-lasting, gains in young children have been produced in those studies in which the families have been made an integral part of the intervention process, especially when this has been done during the children's first three years of life. There is evidence that such gains tend to be long-lasting. Intervention, including family involvement repeatedly and consistently show best results[47].

Parenting training also, probably, affects cultural issues as well. Parents may be led to understand the cultural differences of expectation, importance of education, techniques of motivation, and deal with family structure differences.

On September 12th, 2014, Kristoff and WuDunn wrote in the New York Times Sunday Review, that a Nobel Prize-winning economist, James Heckman, feels that while we should not reduce funds for education, we would be better off were we to spend less on high school and advanced education and prioritize about $1 billion to be spent on parenting and home visitations. In contrast, this author believes that though we live in a wealthy society, which however, certainly has limited funds, we need not reduce educational expenses to increase expenditures to provide parenting training. It is to society's benefit to provide both.

Additionally, according to Kristoff and WuDunn, Heckman feels,

> children's programs are most successful when they leverage the most important — and difficult — job in the world: parenting. Give parents the tools to nurture their child in infancy and the result will be a more self-confident and resilient person for decades to come.

We certainly agree with Heckman's comments; however, there must be

---

[46] Ceci, (1990), p. 49, citing McGillicuddy-Delisi, (1982).
[47] Bronfenbrenner, (1999a), (1999b); cf. Wahlsten, (2002)

better ways to support parenting education than to cut funds for education. Again, we surely agree that of central importance to a child's success is early parenting training, hardly adequately addressed in society. The benefits of parenting may be overwhelming yet cost little. Parenting training is a part of the broad scope of early education.

To the great surprise of many, parenting starts much before a baby is even conceived, and keeps going much beyond when the child has started his or her own family. A couple that I know went to a doctor to see what was wrong because they were disturbed after unsuccessfully trying to have a baby for two or three months. People use various methods to have a baby. Some parents choose to adopt. Some parents losing a pregnancy through natural abortion discover that natural abortion occurs for many would-be parents—a tale hardly ever told but when it happens one learns that it often occurs for many parents.

Also, parenting of an infant is a challenge, but parenting of children becomes increasingly challenging in different ways as a child grows up and matures. I've heard it said, "small children, small problems; big children, big problems." My mother-in-law enjoyed saying getting old is not for sissies. Paraphrasing her, here, I would say, parenting is not for sissies. One father I know left home one morning to go to work on a day just like any other day. He said goodbye to his son as he left the house, never to see his son again. Children leave home in strange ways. I can't imagine the pain one feels the first night, the quite different pain the second night, the first month, and after a year, yet life goes on. The hope and uncertainty are never-ending. The pain of a child dying is quite different. It is final. One reaches closure; of course, not always.

Children get hurt— sometimes just a scratch on the knee, but sometimes much more severely. Parents need to know how to deal with these potentially severe problems. Children do funny things throughout the developmental stages, and some often not so funny, and some do downright stupid things. Parents must be prepared to deal with the variation as these situations appear.

Parenting does not stop when a child leaves the home. Children return, ask for help or need help without asking. When to help and when to refuse help is a difficult problem. Parenting even continues when a child is expecting, and when he or she has a child. Parenting now takes two forms—parenting and grand-parenting. Incidentally, for those who are not yet there, grand-parenting is differently challenging and appears far more rewarding than parenting. As I heard many times, "I wish I could have skipped the parenting stage and jumped right into the grand-parenting stage."

Parenting is made easier if one is aware of the possibilities, if one is prepared for, or simply aware of, likely eventualities. Parenting can be frightening for those who encounter events for which they are unprepared.

Arguably, nothing is more important than parenting (though, of course, not for everyone). For those who seek parenting, it needs to be appreciated every moment, kind of. We all have jobs to attend to, lives to enjoy, and challenges to seek, but when all is said, at the core, for many but not all, life is about children and family. Several doctor-friends have told me, "I never heard a patient on his death bed say, "I wish I had spent more time in the office." And if the patient recovers, this thought gives way to the urgency of the moment nearly immediately. And of course, for many, it must. But in the peace of the moment, the realism sinks in; life for many is family—some might suggest to enjoy it, pursue it, develop it—but continue to do all else keeping everything in perspective. I write this having spent more than a decade of my life, thousands of hours, writing this book. If nothing else, this is a vital message to my own wonderful children, and an apology to my best friend, wife, and precious life partner. Yet while sorry that I did spend that time, never to be recovered, I feel I could not, perhaps should not, have done otherwise. I could have achieved this work quicker, with less procrastination, but that is my internal problem, and, I guess that of those around me as well.

Parenting preparation training programs exist in various parts of the country. For example, a program called Education for Successful Parenting has experienced significant results in parenting training in various high schools in Wake County, North Carolina and Orange County, California. (Education for Successful Parenting, n.d.; Randi Rubenstein, personal communication, June 23, 2013).

It is interesting and revealing that the importance of teaching the sciences, literature, math, and philosophy is well understood and appreciated, but perhaps the most basic human skill and knowledge—parenting—is taught through trial and error, by guess and by gosh, and, all-too-often, the results, evidently, show it. Here, we suggest that parenting is a skill that needs to be better understood, and not merely taught, but learned. Again, all teenagers, all wannabe parents, and new parents need parenting training.

On a related observation, when the mothers of LBW babies receive focused parenting training designed to improve the mother's knowledge, skill, confidence, and self-satisfaction in caring for her LBW infant, the babies' IQs are higher, essentially the same intelligent behavior as NBW babies—an increase of 12.9 IQ points. This remarkable result was achieved with daily sessions each day of the week of discharge from the hospital and with follow-up sessions 3, 14, 30, and 90 days later.[48]

> I am reminded of a student I worked with for about a year, we'll call Jim. He was labeled learning-disabled as were nearly all the children I was asked to help. At the end of the year during which I worked with Jim once or twice a week, he received the most-improved award in his school. While interesting and pertinent to the story, it was the cause of great difficulty. I knew and had commented throughout the year that Jim was hardly learning-disabled, that he could achieve much like most of the other children. His mother knew otherwise. She had been led to believe, when the child was very young, that he would likely have learning difficulties and that she needed to step up to the challenge.
>
> I worked with him, one-on-one, for many hours across the year and knew how quickly and with what ease he comprehended math work and absorbed new information. Yet with each of his successes, minor and major, after I would comment that these successes indicate real potential, the boy's mother would say to me, in front of him, "This is only a glitch; we should not raise his hopes." Jim heard his mother's comments and each time briefly fell back a bit but then quickly rebounded and moved forward. Toward the end of the year, after having had yet another of these interchanges, I am ashamed to say, I told Jim's parents, in private, that if I ever heard her say or heard of her saying those less than motivating words again that I would be out of there. She did, and I never saw Jim again.

---

[48] Achenbach, Phares, Howell, Rauh, & Nurcombe, (1990).

What is important here is that Jim's parents, his mother, would proceed according to a firm preconceived notion from which she could not depart. Incidentally, they may have been right, but I felt the only chance the child had was if they gave him room to succeed. The mother had developed a blind spot that could not be altered, or one which I could not affect. While children should never be given false hopes, where appropriate they should be motivated to reach their potential. Perhaps the mother was right, but I suspect she was convinced early on that she should have no hope. There was reason to reconsider that information. A challenge of parenting is to know when to be optimistic and when not.

I was asked to work with one high school student with spina-bifida. The school and his parents had essentially given up hope for the child. He could not read at age fifteen. But within one semester he read reasonably well and was driven to read more. It was not my tutoring that made the difference, but my patience, interest, and caring. I gave him self-confidence. This resulted in great change in his parents' expectations and how the school dealt with him.

Sometimes teachers, as do all of us, look at a child through a filter. One teacher asked the guidance counselor to get reading help for one of her students. Upon meeting him the first time, I asked him to read something and within a few moments it became clear that he could read just fine. The guidance counselor decided that the problem was that the teacher found the child objectionable and wanted him out of her class for an occasional hour. Teachers are human.

Each of us sees the world through our unique filter that affects how we see people and how we interpret problems. One teacher asked me to focus on math word-problems with a child. (For example, Jane goes to the store to buy four things that cost $1. 42 each. She gives the cashier a ten-dollar bill. How much change should she get back?) The child was fine in the more abstract math problems, adding, subtracting, and multiplying numbers, but, according to the teacher, could not deal with word-problems. Within minutes, it became clear the child had no difficulty with word-problems; he had problems reading. When I read the problems to the child he had no difficulty. The teacher assumed the problem to be a category in the curriculum, word-problems, which they had just started. Incidentally, if the teacher had had time to work with the child one-on-one, he would have noticed the cause of the problem, but with fifty or sixty eyes looking around the classroom, at the same time, teachers often cannot have the time to recognize the real problems. But, more to the point, why was the teacher unaware of the child's reading problem? First, he may have been aware but did not put the two issues together. Second, classes are often team taught; one teacher teaches reading and other language subjects, while another teacher focuses on math and science. If the teachers do not compare notes, problems fall through the cracks.

Assuring adequate early education (preschool) experiences

Nicholas Kristof wrote an editorial which appeared in the New York Times on November 19, 2014. With pointed humor he said,

> Indeed, we love [our children] so much that, on average, child care workers earn almost as much per hour ($10.33) as workers who care for animals ($10.82), according to a new study from the University of California, Berkeley.
> [Turning to preschool attendance, he writes,] We love them so much that only 38 percent of American 3-year-olds are enrolled in education programs. The average is 70 percent among the 34 industrialized countries in the Organization for Economic Cooperation and Development.

Note that the 38 percent of American three-year-olds in preschool programs are likely advantaged children. The disadvantaged three-year-old children likely receive essentially no preschool education. This needs change.

Reporting on 2000 data, in another report focusing on four-year-olds, Magnuson and Waldfogel (2005) wrote that 60 percent of young children attend preschool in the year before they enter kindergarten. So, 38 percent of three-year-olds, moving to 60 percent of four-year-olds attend preschool in the United States while, in the rest of the Western World, a much larger percentage of children receive early education.

Researchers have reported a wide discrepancy in the value of preschool attendance. With what sometimes appears the use of smoke-and-mirrors, researchers focus their analyses on different aspects of preschool benefits. Their arguments appear to pass each other unnoticed like ships on a foggy night. This needs understanding and clarification.

In 1994, Herrnstein and Murray (p. 405) indicated that preschool education results, initially, in an average increase of seven points in IQ, but this advantage drops to an average of three points by third grade[49]. Many researchers have asked and probed, why this drop occurs. No one has produced direct evidence explaining this. The cause of the drop is likely environmental. It may well be related to problems stemming from economic disadvantage. And some of this drop may well have been reduced by adding breakfast to the free or reduced-price *lunch program* since these findings were first reported.

Some researchers suggest there is no lasting IQ increment from preschool education. For example, Jensen reported in 1969 that

> Most positive claims for the efficacy of Head Start involve evidence of the detection and correction of medical disabilities in disadvantaged preschool children and the reportedly favorable effects of the program on children's self-confidence, motivations, and attitudes toward school. (p. 3)

Jensen made a similar judgment in 1998, thirty years later:

> The general conclusion of the hundreds of studies based on Head Start data is that the program has little, if any, effect on IQ or scholastic achievement that endures beyond more than two to three years after exposure to Head Start. The program... does, however, have some potential health benefits, such as inoculation of enrollees against common childhood diseases and improved nutrition (by

---

[49] See also Currie, 2001, pp. 217-226

school-provided breakfast or lunch). The documented behavioral effects are less retention in-grade and lower dropout rates. (p. 338)

Some might conclude that, if the accomplishments Jensen reports are
all Head Start does, it's a great success. Furthermore, it has recently been
indicated by Nisbett and associates in 2012 that

the discrepancy between school achievement effects and
IQ effects (after early elementary school) is sufficiently
great to suggest to some that the achievement effects are
achieved more by attention, self-control, and perseverance
gains than by intellectual gains per se (Heckman, 2011;
Knudsen, Heckman, Cameron, & Shonkoff, 2006, p. 138).

Nisbett explains, in 2005, that while Head Start shows this decrement
in IQ growth by third grade,

more ambitious interventions produce very significant
gains that last as long as until age 15, the oldest age tested
to this point to my [, Nisbett's,] knowledge (S. L. Ramey &
Ramey, 1999). ... [It] is not merely early intervention that
increases IQ and school achievement. Programs at every
age level from infancy to college can be effective (Bennett,
1987; Herrnstein, Nickerson, De Sanchez, & Swets, 1986;
Selvin, 1992; Steele et al., 2004; Treisman, 1992). (pp.
303, 304)

The real objective to be achieved is not merely attending preschool programs but attending quality preschool programs. And as Nisbett argues, this
may be expanded to include quality education at all ages.

A related study, in 2015, by Muschkin, Ladd, and Dodge, of Duke
University, reported on

the community-wide effects of investments in two early
childhood initiatives in North Carolina (Smart Start and

More at Four) on the likelihood of a student being placed into special education.

They found that attendance in "both programs significantly reduce the likelihood of special education placement in the third grade, resulting in considerable cost savings to the state." Furthermore, they added that the effects of the, "two programs differ across categories of disability, but do not vary significantly across subgroups of children identified by race, ethnicity, and maternal education levels." They conclude,

> At the 2009 funding levels, the SS [Smart Start] reduces [special education] placements by 10% and the MAF [More at Four] program by 32%. Together, at these funding levels, the two programs reduce the odds of special education placement by 39%.

That is just one justification for the widespread introduction of early education. Note, importantly, that special education services cost 40 percent more than regular class services. This suggests pre-school education results in large cost savings.

This also, less than subtly, conveys that the identification of the need for special education services is less than scientifically determined. The preschool programs should not affect identification of special education services need.

On December 19th, 2014, Joel Klein, a previous Chancellor of Education in New York City said about early education,

> We're starting kids way too late, particular kids in a challenged environment. The thing that bothered me most when I was Chancellor, [was that] the average child from a poor community started school with about 20% of the vocabulary of kids from a middle-class community. That's a huge gap, and that gap grows [as a child moves through his education].

That discrepancy in vocabulary reflects a significant difference in home environment between the advantaged and disadvantaged, one that can be substantially remedied by early education and by parenting education.

To expand on the work mentioned earlier in the section on parenting training, Heckman co-authored several studies, at least two of which were published recently in *Science*. One of them entitled, *Early Childhood Investments Substantially Boost Adult Health*, and whose primary researcher was F. A. Campbell, concluded "Children [who had been enrolled in a very early education program] ... have a significantly lower prevalence of risk factors for cardiovascular and metabolic diseases in their mid-thirties." (p. 1483). Many fewer of the participants were obese, and they were more likely to have medical insurance.

The second study entitled, *Labor market returns to an early childhood stimulation intervention in Jamaica,* whose primary researcher was Paul Gertler, consisted of participants who had been growth-stunted toddlers. These children were divided into two subgroups; some were assigned to a control group and the other group, let's call group B, received two visits every week from a health aide who counseled the parents on how to parent. Twenty years later, the individuals of group B were not only less likely to have committed violent crimes, but also remained in school longer, and additionally were earning 25 percent more than those individuals who had been assigned to the control group. Much like the author of this book, the researchers concluded that very early education in parallel with parenting education is a powerful intervention.

To be clear, early education encompasses not merely starting education at the toddler age, but also parenting education starting, at least, during prenatal care and continuing the parenting education up to when the child enters kindergarten. It includes conveying to pregnant women the risks of smoking, drinking, taking drugs, and the risks of toxins such as lead paint. The costs are not trivial but far less than the likely expenses for additional medical care and costs for many problems of society.

Part of this thinking should be that early education, that is, from infancy to kindergarten, would be most effective when adequate nutrition and health processes are provided. This, too, was shown in the Jamaica study referenced just above.

The impact of early schooling on intelligence test results is strikingly demonstrated in the "natural experiment" created by the custom of enrolling new students in primary school once each calendar year, and of forbidding enrollment to children who have not reached a required birthday, so that children born just before the cutoff date start school a year ahead of children born just after it. Researchers have taken advantage of this pattern to attempt to independently correlate test results with chronological age and length of schooling. In the path-breaking 1969 study that introduced this approach, later summarized by Ceci in 1991 (p. 708), German researchers Baltes and Reinert "randomly sampled 630 children from 48 elementary schools," identifying "three cross sections of 8- to 10-year-olds" who were "separated in age by four-month intervals" to be "administered a German version of the Primary Mental Abilities Test (Thurstone and Thurston, 1962)." Baltes and Reinert

> found a substantial correlation between the length of schooling and intellectual performance among same-aged, same SES children. Highly schooled 8-year-olds actually were closer in mental abilities to the least schooled 10-year-olds than they were to the least-schooled 8-year-olds.[50]

Collectively, these studies suggest that preschool and early education significantly enhance intelligent behavior, and if made available for all would reduce the IQ gap between the economically disadvantaged and advantaged. Notably, many of these studies ignore race.

It is recommended that preschool, quality preschool, be made accessible to *all* children, especially the disadvantaged, because the advantaged already attend preschool in greater numbers, and moreover, because they tend to be enrolled in quality preschool.

Assuring adequate formal and informal learning experiences in summers

Formal and informal learning experiences during school breaks, the longest of which currently occur in summers in the United States, are often available

---

[50] Ceci, (1991), p. 708, reporting on Baltes and Reinert, (1969); cf. Ceci and Williams, (1997).

only to children of economically advantaged families. Writing in 1991, Ceci (p. 705) explains that

> a clear illustration of the influence of schooling on IQ performance is seen in the small but reliable decrement in IQ that occurs during summer vacation, especially among low-income youngsters whose summer activities are least likely to resemble those found in school (Jencks, et al., 1972). This finding has been replicated twice with larger samples by independent investigators (Hayes [& Grether], 1982; Heyns, 1978).

And in 2012, Nisbett and associates (p. 138) elaborate,

> the knowledge and skills of children in the upper fifth of family SES, actually increase over the summer (Burkam, Ready, Lee, & Lo Gerfo, 2004; Cooper, Charlton, Valentine, & Muhlenbruck, 2000), an effect that is likely due to enriched activities for the higher SES children. This effect is so marked that by late elementary school, much of the difference in academic skills between lower and upper SES children may be due to the loss of skills over the summer for lower SES children versus gains for higher SES children (K. L. Alexander, Entwisle, & Olson, 2001).

These researchers attribute the relative gains of richer over poorer children to better opportunities for formal and informal learning during school breaks. Additionally, gains are also partly due to the adequate nutrition children receive while participating in those richer activities. Note that disadvantaged children do not receive free or reduced-price breakfast and lunch during school breaks.

So, it is recommended that 1) formal and informal learning experiences during school breaks be provided all children from preschool through high school, and that 2) adequate nutrition be provided all children, throughout the year as a foundation for learning. There is reason to believe that just providing additional nutrition during these school breaks would prevent at least some of the drop in IQ experienced during school breaks.

Effective formal and informal learning experiences, kindergarten through high school

Formal and informal learning experiences result in the accumulation of information and strategies, which in-turn enhance one's intelligent behavior. For good reason, a child learns addition before learning multiplication. The stages of learning math, for the most part, build on each other, and this is true for many areas of learning. The classic *Huckleberry Finn*, read at age eight, appears to be a different book from *Huckleberry Finn* read in high school, and different again from *Huckleberry Finn* read in college. The book does not change over time; the reader's mind does.

Age-appropriate intelligent behavior, which appears as a stable IQ, depends on a steady growth of intelligent behavior: the more experiences, the more knowledge and strategies to support enhanced intelligent behavior. The growth of intelligent behavior depends on cognitive and memory facilities, adequate nutrition and health, and on the accumulation of formal and informal learning experiences. A healthy child that is adequately nourished will more likely maintain an age-appropriate growth of intelligent behavior and thus an expected IQ; a less-than-well-nourished child will likely absorb less new information, and for that reason alone, fall increasingly behind academically, slip in the age-appropriate growth of intelligent behavior, and accordingly his or her IQ will likely steadily and increasingly drop.

In 2012, Brinch & Galloway (p. 425) found that additional years of education in the middle teenage years have "a substantial effect on IQ scores measured at the age of 19". Several researchers[51] have shown that for every year a student drops out of high school, he or she loses about 1.8 points in IQ. Some children drop out of school gradually and subtly starting in the early grades for a variety of reasons—family problems, cultural orientations, psychological or emotional issues, limited nutrition, or, more broadly, lack of resources resulting in limited well-being.

Scholarship on secular improvements in intelligent behavior has challenged this interpretation—the "nutrition hypothesis"—with the hypothesis

---

[51] For example, Härnqvist, (1968a), (1968b); Husen, (1951); DeGroot, (1951).

that economically disadvantaged children receive less cognitive stimulation. Colom, Lluis-Font, and Andrés-Pueyo (2005) argue that generational gains in intelligent behavior between two samples of seven-year-old boys of identifiable socio-economic status tested in 1970 and 1999[52] were "mainly concentrated in the bottom and medium halves of the distribution" and "the gains are progressively smaller as we move toward the upper half of the distribution" (p. 87) and argue that "nutrition and health care" explain much of the improvement. Marks in 2010 (p. 643) makes a similar argument but proposes in addition that "secular and racial differences in IQ are an artifact of variation in literacy skills" rather than directly and indirectly the result of poor nutrition or the result of general lack of cognitive stimulation; that is, that the problem is inhibited opportunities to learn to read among the economically disadvantaged.

Additionally, several authors, including Herrnstein and Murray (p. 400) reported in 1994 that years of research indicate that supplemental learning enhances performance on SAT tests. Sternberg, in 1996, and others reported, similarly, that supplemental learning enhances IQ scores, and, more generally, that IQ is affected by formal and informal learning. Nearly three decades ago, in 1990, Ceci wrote that

> numerous training projects [had] shown IQ to be fairly malleable, even showing large gains over brief training periods (e.g., Feurstein, 1979; Gardner, 1987; Herrnstein, et al., 1986). Of special interest has been the finding that the largest IQ gains in intervention in a high risk sample are found among those families at greatest risk (Ramey & Ramey, 1989.) Feurstein et al., (1979) specifically addressed retarded [IQ of 70 or below] adolescents. The central observation throughout is that environmental enhancements, not strangely, advantage [benefit] the disadvantaged most. Gardner (1987) advances methods to recognize and help individuals achieve their unique intelligence. (p. 139).

---

[52] Colom et al. (2005) collected their data in Barcelona, Spain. The mean gain in twenty-eight years was 9.7 IQ points (p. 86).

In 1986, Herrnstein, Nickerson, DeSanche, and Swets reported that in one educational study in Venezuela, nine hundred seventh-grade students received additional training in intellectual activities such as visual-spatial and verbal reasoning, as well as vocabulary and word analogies; those in the experimental group who received sixty additional forty-five minute lessons during the course of a year increased their IQ scores between 1.6 and 6.5 IQ points over those who did not receive the extra training. A few years later, in 1994, Herrnstein and Murray (p. 400) concluded that on

> the SAT, partly an intelligence test, … 45 hours of studying adds about .16 standard deviation (about 18 points) to the verbal score and about .23 standard deviation (about 28 points) to the math score.

Studies comparing high school dropouts with high school graduates, controlling for IQ, SES, and school performance before high school show a loss of IQ related to the missed schooling[53]. Sternberg, in 1996 (p. 117), wrote, evidence of this type supports the claim that "the single best predictor of adult IQ is not parental IQ or income level or social class or any of the variables one might expect. Instead, it is number of years of schooling and especially Western schooling."

I would extend that assessment: what matters is years of formal and informal learning experiences, *while being adequately nourished, or, increasing the scope, having adequate well-being.*

The view taken in this book combines the nutritional and cognitive stimulation hypotheses: intelligent behavior is most likely to grow age-appropriately if a child receives sufficient nutrition and care while experiencing an adequate continuous education. With adequate nutrition and well-being, formal and informal learning experiences and supplemental training activities more effectively enhance intelligent behavior.

Formal and informal learning is more effective when adequate nutrition and good health are maintained. Evaluating the effects on intelligent behavior of formal and informal learning experiences assume that individuals are adequately nourished and healthy through the developmental stages so that

---

[53] Ceci, (1991); DeGroot, (1951); Härnqvist, (1968a,) (1968b); Husen, (1951)

their intelligent behavior can grow age-appropriately through accumulating information and strategies.

The various benefits of formal and informal learning (including pre-school experiences, parenting training, the thirteen school years starting with kindergarten through high school, and summer experiences) are all separable from the previously discussed considerations of the earliest developmental stages. Again, adequate nutrition and well-being is required to achieve the expected benefits from the array of formal and informal learning experiences.

Incidentally, improving learning experiences before enhancing nutrition likely presents limited benefits.[54] Also, except for extended summer school benefits and early parenting training, there has been no mention of enhancing schooling. Current school educational programs are assumed. Certainly, several studies have shown that change in educational processes could significantly benefit the disadvantaged, but this is not considered here.

To achieve the benefit of formal and informal learning experiences, it is recommended that a year-round, steady adequate diet, including age-appropriate standard pediatric multivitamin supplements when needed, be made available to all children to age eighteen.

The extent to which formal and informal learning experiences are cumulative

Although maintenance of the capability for momentary intelligent behavior is necessary to facilitate the growth of intelligent behavior, maintenance of intelligence and growth of intelligence must be distinguished. Maintenance of underlying, genetically determined cognitive and memory facilities has been discussed for the initial developmental stages up to age four. The growth of intelligent behavior through formal and informal learning experiences during the school years is analyzed here *assuming* maintenance of the underlying genetically determined cognitive and memory facilities through adequate nutrition and health (adequate well-being).

IQ scores are dependent on age-appropriate formal and informal learning and on overcoming environmental constraints that interfere with

---

[54] In deprived areas of the world, it seems appropriate that one should assure adequate nutrition before, or at least in parallel with, providing new schools.

learning. As reported, after a student drops out of high school, he or she loses about 1.8 points in IQ *per year* of missed schooling[55]. The effects of formal and informal learning are cumulative, so the effect of being prevented from effectively benefitting from formal and informal learning experiences in and out of school is also cumulative. Besides dropping out, children may be absent due to human or natural catastrophes. Those who attend school while less than well-nourished are less able to benefit from learning experiences, and thus, in fact, to some degree, miss school.

Surely, information gained from formal and informal learning experiences is not *entirely* cumulative, but it is reasonable to acknowledge that the benefits of formal and informal learning experiences are known to be significantly cumulative. IQ test scoring assumes age-appropriate increases in strategies and knowledge. Economically disadvantaged children are less likely to be well nourished, and, accordingly, less able to benefit from formal and informal education, and are thus likely to steadily lose ground in IQ scores. And if economic disadvantage continues over a span of years, then IQ scores are likely to steadily and gradually decrease over that period.

The loss of IQ during summer breaks, especially of the economically disadvantaged, is independent of the other effects on intelligent behavior during the regular K–12 school years (see Ceci, 1991). So, this effect is viewed as cumulative with the effects of age-appropriate growth of intelligent behavior during the K–12 years.

Parenting training also has independent effects on a child's intelligent behavior in that this affects the child's intelligent behavior indirectly; they affect motivation, expectations, adequate nutrition, and the teaching methods of the parent by, for example, suggesting showing the child how to do something rather than doing it for him or her. Also, pre-school experiences relate to a period of formal and informal learning prior to K–12. All these aspects relating to intelligent behavior deal with formal and informal learning experiences.

Admittedly, all the studies cited relate to separate phenomena. There are no studies of all these factors together that show unequivocally that their combined effects are cumulative. However, the effects of summer experiences and preschool benefits have been shown to be independent of the

---

[55] Ceci, (1990), p. 77; Härnqvist, (1968a), (1968b)

regular K-12 formal and informal learning experiences. Furthermore, each intervention discussed occurs in a distinct period in a child's life. Therefore, arguably, the impact of these formal and informal learning interventions on intelligent behavior is cumulative.

To be sure, overlap of learning is experienced throughout these years. Awareness of the potential for overlap is reflected in the use of lower, more conservative, values from the range of effect reported for each intervention. While it would be questionable to claim these values to be unequivocal, it appears reasonable to claim that significant benefit is gained from each of these interventions, and that the benefits discussed are largely cumulative. Specific values are addressed in the next chapter.

Early environmental effects are independent and cumulative with later learning experiences

Formal and informal learning experiences, assuming continued well-being, build on the early nutritional and birthing considerations. Nutrition influences one's well-being, determining current intelligent behavior. Formal and informal learning experiences effect *growth* of intelligent behavior, resulting in changes in content as well as connectivity within the neuronal system. So, our discussion of nutrition and health addressed one aspect of intelligent behavior, intelligent behavior in the moment, while our discussion of formal and informal learning experiences focused on *age-appropriate growth* of intelligent behavior, a different aspect.

Mothers influence intelligence differently from fathers

As has already been shown, mothers have considerably more influence over the intelligence of their offspring than do fathers. This should not be surprising in that mothers, alone, provide prenatal care and often provide most perinatal care as well.

Additional direct evidence is also available. According to the Reed and Reed study, in 1965, when a mentally retarded male mates with a normal IQ female 7.8 percent of the resultant offspring are retarded, whereas when a mentally retarded female mates with a normal IQ male 19.6 percent of the resultant offspring are retarded. This indicates that retardation is two and half times more likely for the offspring of a retarded female and a normal IQ male than a retarded male and a normal IQ female. This strongly suggests the overwhelming importance of the mother's intelligence in determining, and influencing, the offspring's intelligence. To be clear, this has nothing to do with genetic considerations because in one case the father was mentally retarded and in the other the mother was mentally retarded. However, the results show environmental difference; the normal mother has a much stronger positive influence on the child's IQ than the mentally retarded mother.

In 1972, Jensen (p. 220) reasonably suggests two hypotheses to explain this. First, retarded mothers appear to provide an early environment severely lacking in the kinds of mother-child interaction that promote mental development, and second, mentally retarded mothers may provide poor prenatal environment for the developing fetus. The mother whose IQ is 70 or less is far more influential than the father whose IQ is 70 or below, in determining the offspring's low IQ. While these are appropriate hypotheses, mention needs to be made of the more powerful significance of the normal IQ female's influence on the offspring. Again, the influence is environmental. Another comment should be reiterated here; mental retardation may be entirely caused by inhibited environmental factors – not genetic factors. The IQ of mentally retarded mothers and/or fathers may be enhanced with enhanced environmental factors. Of course, this would not affect Jensen's above-mentioned hypotheses.

A massive research effort run by the National Institute of Health collected data about the offspring of 42,000 women. Willerman and associates, studied, in 1974, a subset of the data—interracial unions. They identified 129 interracial children, one group of whom had white mothers while the other had black mothers. Willerman and associates found that at age four the interracial children of the white mothers had a mean IQ of about nine points above that of the children of black mothers. There is no reason to suggest that racial genetic factors caused this difference because the mates of the black mothers were white, and the mates of the white mothers were black. Furthermore, it conveys that mothers have more influence on their children's intelligence than fathers. *Perhaps it needs to be highlighted that while there is no evidence that race plays a role in determining intelligence, there is overwhelming evidence that being disadvantaged does influence intelligence, and blacks are disproportionately disadvantaged.*

The results of these studies argue persuasively that the mother's influence on offspring's aptitude is greater than that of the father's, at least regarding prenatal and perinatal care. The same seems to be the case in later stages of development. In 1981, Scarr (p. 16) reported a study of maternal effects[56] (Willerman, 1974) that showed that college students whose mothers are more highly educated than their fathers have higher aptitude scores than those whose fathers are more highly educated than their mothers.

Smilansky Study: a long-term study

One fascinating and revealing study took place in Israeli kibbutzim. It not only addresses that intelligence can be significantly enhanced among different ethnic groups in Israel, but it also indirectly addresses the black-white gap in intelligence in the United States. To summarize the description given by Scarr, in 1981 (p. 51), Smilansky reported on the IQ scores of 1,340 children in 129 Israeli kibbutzim; half of the children were Ashkenazy (of European Jewish parents), the other half being Sephardic (of oriental Jewish parents). Both groups of these children were reared together, first in communal nurseries and then, through adolescence, in small groups with their caretakers. Though they visited their parents daily for two hours, their education, nourishment, and health were uniformly determined entirely by and within their communal setting; they resided in the children's groups.

Interestingly, home-reared Sephardic children have, on average, an IQ of 92, while home-reared Ashkenazy children score, on average, sixteen points higher at 108. (Note that the difference in IQ, 16 points, was about the same as the IQ gap between blacks and whites in the United States at that time.) These communally-reared pairs of children, one of Sephardic and the other of Ashkenazy parents, were tested periodically from age five to fourteen and had equivalently high scores that averaged about 115. The Smilanskys matched each of 670 Oriental children with a European child within the same kibbutz, controlling for parental educational level, length of residence in Israel, and several other factors. As Scarr documented, the children were tested with the Stanford-Binet (four to five-year-olds) or the Wechsler Intelligence Scale for Children (six to fourteen).[57]

---

[56] Willerman and associates, (1974)
[57] Scarr, S. 1981 p. 51, personnel communication, June 12, 1973

Several observations may be made. First, intelligence, as measured by IQ tests, can be substantially enhanced by environmental conditions. The IQ averages of these two groups were raised essentially to the same level though the scores of the Sephardic children were raised by an average of twenty-three points while that of the Ashkenazy children were raised by seven points. The initial difference of sixteen points may have resulted from different prenatal and perinatal care. Second, and importantly, this study suggests that inferior early conditions may be overcome by superior subsequent conditions. Third, those with relatively greater disadvantaged histories are more advantaged by subsequent benefits. Fourth, IQ averages may be considerably higher than generally believed. Furthermore, the range of intelligence, given appropriate environmental conditions, is likely narrower yet higher than currently believed; that is, the 95 percent range might well move from the currently defined 70 to 130, to about 90 to 135. This relates immediately to the observations made in Chapter 2 that intelligence of the bulk of the population, 95 percent, could have a narrower range with a higher mean. Finally, also discussed in Chapter 2, it is likely that there may be considerably fewer mentally challenged people than currently believed.

In net,

1. Nutrition and health have powerful influence on both intelligence in the moment, as well as, growth of intelligence—prenatally and post-natally.
2. Providing mother's milk to babies has significant effect on subsequently measured intelligence.
3. Lowering frequency of LBW baby births significantly increases average IQs.
4. Parenting training should be promoted in high school. These courses should cover the issues discussed in this book. Teens need to be aware of what to expect and what will be expected of them as parents. Parenting training should be provided potential parents before pregnancy is expected.
5. Very early quality education, starting for three-year-old children, has immense effect on intelligence, especially among the disadvantaged.

6. Provide educational opportunities during school breaks.
7. Provide nutritional supplements during school breaks.
8. Mothers have more influence on their children's intelligence than do fathers.
9. The two programs in North Carolina, Smart Start and More at Four, "reduce odds of special education placement by 39 percent. Note that special education services cost 40 percent more than regular class services.

CHAPTER 5

The IQ Gap between racial groups, as well as between the advantaged and disadvantaged, can be essentially eliminated.

The threads of the book's tapestry introduced in this chapter may appear as white and black. Symbolically, and as may be recalled from physics, the white threads actually consist of all the colors of the spectrum. Interestingly, the threads of this chapter do not convey an appearance of white or black, but of advantaged and disadvantaged. This may present confusion in the tapestry, but the message is straight-forward; the IQ gap between blacks and whites is really an IQ gap between a group of people who are predominantly disadvantaged and a group of people who are predominantly advantaged; the issue discussed proves to be not racial but economic status.

While doing this research I called the education director in Bermuda to ask for their data on the difference in IQ between black and white children. He was dismissive of my request saying, "We have no such data." Interestingly, we, in the United States have little data focused on the IQ gap between the advantaged and disadvantaged; our data is predominantly racial, about the IQ gap between blacks and whites. This dichotomy should give one pause.

So, there is difficulty in discussing directly the IQ gap between the advantaged and disadvantaged. However, while data about the IQ gap between the disadvantaged and advantaged is rather sketchy, there is overwhelming data about the IQ gap between blacks and whites. Currently, there are data describing differences in IQ between the races (whatever race might mean).

The average IQ score of blacks is considerably lower than that of whites, which is lower than that of Asians. The question is not whether this is true but why this is true. The question is, are these differences in IQ caused by genetic differences, as some researchers have claimed, or are these differences caused by nutritional, health, motivation, expectations and other environmental differences – differences over which individuals and society have significant control?

Because essentially all researchers believe that an individual's intelligence is at least partly determined by genetic factors, it follows, for many, that the IQ gap between blacks and whites is also at least partly determined by genetic factors. In their 1994 book, Herrnstein and Murray wrote, "It seems highly likely to us that both genes and the environment have something to do with racial differences [in IQ]." While this certainly appears rational and reasonable at first glance, it does not stand up to scrutiny. This chapter was written to refute this notion. Though there is no doubt that an *individual's* intelligence is partly determined by genetic factors, it does not follow that differences between *groups of individuals* is determined by genetic factors; it does not follow that the IQ gap itself, the IQ gap between blacks and whites, is even partly determined by genetic factors. To repeat differently for emphasis, the gap between the races could be entirely determined by environmental factors. While genetic factors do influence the intelligence of individuals, there is no reason to assume they determine differences in intelligence between groups of individuals though, of course, they could.

Let me explain with an analogy. If the average IQ of a group of one hundred *disadvantaged white* students was compared to that of a group of one hundred *advantaged white* students (again, all being white), one would expect to find, all other variables being controlled, that the disadvantaged white students' IQ scores be considerably lower than that of the advantaged white students. The data would show an advantaged/disadvantaged IQ gap. Research published by Loehlin and associates shows precisely this. Though, surely, differences in IQ between individuals are partially genetically determined, one should *not assume the differences between these groups,* disadvantaged white students and advantaged white students, to be even partially genetically caused. Rather, one might reasonably assume the differences in IQ to have been caused by differences in advantages, including differences in nutrition, health, experiences, and other environmental differences.

However, when discussing the IQ gap between the blacks and whites many feel it reasonable to assume that genetic differences must, at least, partially explain the IQ gap. Why assume the black/white IQ gap itself is partly genetically determined, when we know that in the United States 8 percent of whites live under the poverty line while 25 percent of blacks live under the poverty line. Isn't it at least possible that the black/white IQ gap, again the gap itself, is entirely caused by the environmental differences experienced by the two groups? This was the findings for the study of white advantaged and white disadvantaged students. Why wouldn't this apply similarly to black and white students who are known to be affected by differing advantages? Shouldn't the environmental differences be examined to determine their contribution to the gap, and if environmental differences between the groups can be shown to entirely explain the black/white IQ gap, then wouldn't it be shown that genetics need not play any role in explaining the gap? Precisely that will be shown in this chapter. Incidentally, no genetic differences in intelligence have ever been identified between any racial groups.

Scientific integrity demands that I acknowledge that if, in the future, research reveals that genetic factors play some role in determining the IQ gap between blacks and whites, then this statement is wrong, but until such time that that is demonstrated we feel it safe to say that if environmental differences between the groups, any groups, is shown to entirely explain the IQ gap between those groups, then it is shown that genetics need not play any role in explaining the gap.

To highlight this reasoning with yet another analogy, if a child riding a bike falls and hurts her head such that she becomes mentally challenged for life, one does not proclaim the child's mental difficulties to be partly genetically determined, though, of course, it could be. One more reasonably concludes the mental challenge is entirely caused by the fall. Likewise, if an individual is disadvantaged and therefore experiences limited nutrition and poor health, one should not proclaim that his or her limited intelligence is partly due to genetic factors. But rather, one should more reasonably determine whether environmental factors explain the limited intelligence, and if so, conclude that genetics need not play any role.

A theoretical study helps further explain. Kamin wrote,

A simple hypothetical but realistic example shows how the heritability of a trait within a population is unconnected to the causes of differences between populations. Suppose one takes from a sack of open-pollinated corn two handfuls of seed. There will be a good deal of genetic variation between seeds in each handful, but the seeds in one's left hand are on the average no different from those in one's right. One handful of seeds is planted in washed sand with an artificial plant growth solution added to it. The other handful is planted in a similar bed, but with half the necessary nitrogen left out. When the seeds have germinated and grown, the seedlings in each plot are measured, and it is found that there is some variation in height of seedling from plant to plant within each plot. This variation within plots is entirely genetic because the environment was carefully controlled to be MZ [essentially uniform?] for all the seeds within each plot. The variation in height [within each plot] is then 100 percent heritable. But if we compare the two plots, we will find that [on average] all the seedlings in the second are much taller than those in the first. This difference is not at all genetic but is a consequence of the difference in nitrogen level. So, the heritability of a trait within populations can be 100 percent, but the cause of the difference between populations can be entirely environmental. (p. 118)

This hypothetical analogy shows that while there are genetic differences in intelligence among humans, the differences in intelligence among two groups of individuals may be entirely determined by environmental differences.

Surely, we are all genetically different in every possible way, but nutrition, health, experiences, and other environmental factors influence the expression of genetic factors. The intelligence of each one of us is partially genetically determined, but the expression of those genetic influences is affected by environmental differences. If one were to randomly select a thousand black children and provide them with inadequate nutrition, their average IQ would be lower than an equal number of randomly selected white children provided adequate nutrition. The IQ gap between the groups may have nothing to do with genetic differences between the groups.

Certainly, the races are different genetically. That is not in dispute. Over the many millenniums blacks have evolved different characteristics, some differences being quite recognizable; kinky hair is an example. But these are merely differences. Every one of the readers of this book is different, but that does not mean that the group with brown hair is measurably different in their spectrum of intelligence from those with red hair, or, as in my case, little hair. There is *no* evidence to suggest that genetically determined hair color, or hair texture, or race affects intelligence differences among groups of people. Similarly, there is no evidence that *genetic factors,* genetic differences, cause the disadvantaged group of readers of this book to have lower intelligence than the advantaged group. More to the point, there is *no evidence* that genetically determined racial differences affect intelligence. This statement has been made earlier in this book, yet scientific integrity demands that I also observe that "absence of evidence is not evidence of absence"[58]. But I can say that I have never seen any evidence that shows genetic factors influence intelligence between any racially different groups of people. Yet, if some evidence does surface, then this observation would have to be reconsidered.

While average IQ scores among the members of the races are different, to claim that differences in intelligence are due to genetic differences, one must know of genetic differences that cause differences in intelligence. Short of that, one must first reasonably discount non-genetic explanations of the differences. One cannot claim that the black/white IQ gap is even partly determined by genetic factors, as many researchers claim, until one has definitively dismissed all other potential causes, especially when one recognizes the wide disparity among the races in environmental conditions. And, even more especially, when one realizes that no genetic differences in intelligence between racially different groups of people, including blacks and whites, have ever surfaced.

I have tutored and mentored many more black children than white. Blacks are much more likely to be disadvantaged and thus environmentally constrained and thus less nourished and have limited well-being. There are

---

[58] As indicated in personal communications by Henry Shaffer, a friend and geneticist at North Carolina State University.

other explanations in play such as motivation, self-confidence, expectations from teachers and parents and I will add society, as well as self-image. The importance of expectations can hardly be overestimated.

Pertaining to the last explanation, I have heard several students, one as young as seven years old and others in middle school tell me, "I am black. I am stupid." I have also heard, "I am a girl. I can't do math." How does one respond to such comments? Incidentally, these children said these words without humor, without emotion, without analysis—as just plain fact. Of course, they had heard these comments all around them. They may have been told this directly by one or another person or even by several. And if heard often enough, or even just from one significant person, one does not doubt or question these words. They come to believe, I am stupid because I am black, or, I am a girl and therefore can't do math, and, of course, for many, they come to believe there is nothing that can be done about that.

This chapter is written to make clear that the claim that differences in intelligence among races is partly genetic has no credibility.

Specifically, while the IQ gap of fifteen points between blacks and whites has recently been decreasing, a significant gap remains. This is hardly in dispute. In particular, until about the 1980s, blacks scored an average of about 85 on IQ tests while whites scored an average of about 100. The question is, is the gap caused by genetic factors or environmental factors? Some still claim the answer is both— that difference in intelligence must be partially caused by genetic factors. Sure, they say, one can reduce some of the IQ differences through environmental enhancements, but, they claim that that part is relatively trivial, and that that part that is genetically determined cannot, currently, be changed.

In contrast, the explanation for the black/white IQ gap may well be that blacks, on average, have been and continue to be disproportionately disadvantaged. With this awareness, and with select environmental interventions, we can finally claim that the circular pattern, the central issue discussed in this book, can finally be broken. More particularly, we know how to break the circular pattern that *blacks, as a group, are disproportionately disadvantaged because they show limited intelligent behavior,* and, *they show limited intelligent behavior because they are disproportionately disadvantaged.*

Toward the middle of the last century, Loehlin revealingly wrote,

Questions concerning the meaningfulness and predictive utility of any estimates of general intelligence, the stability of such estimates, and the relative contributions of genetic and environmental factors to intelligence have remained among the most difficult and emotionally charged issues within the social sciences for more than five decades [now more than eight decades]. When these questions are reexamined in the context of racial and social class differences in a society ridden with unresolved tensions in these areas, it is not surprising that the result should be a massive polemic in which personal conviction and emotional commitment often have been more prominent than evidence or careful reasoning. (pp. 2-3)

While Loehlin's comments are certainly true and are essentially just as valid today as when written, many researchers involved with these discussions have carefully considered views, sometimes strong views, that do not agree with that of others. Evidence used may certainly be in error, or there might be some evidence that counterbalances the importance of other evidence. These are the legitimate discussions. Others are irrelevant. Any scientific discussion, whether supportive or in conflict with the conclusions indicated here, is welcome. The objective is to reveal truths.

Jensen, and many other researchers including Herrnstein, Murray, Lynn, Rushton, and Eysenck, too many to list, documented their findings conveying that blacks have lower IQs than whites, largely, or at least significantly, they claim, because of genetic factors. Though they understand that IQs of blacks is lower than that of whites partly due to environmental differences, they claim that IQs of blacks is lower than that of whites, at least significantly, because of genetic reasons. And they hold fast to this belief because, they claim, both genetic and environmental factors influence intelligence of individuals.

According to Loehlin (p. 8), Jensen rather strongly writes,

So, all we are left with are various lines of evidence, no one of which is definitive alone, but which, viewed all together, make it a not unreasonable hypothesis that genetic factors

129

[Jensen claims] are strongly implicated in the average Negro-white intelligence difference. The preponderance of the evidence is, in my opinion, [Jensen writes,] less consistent with a strictly environmental hypothesis than with a genetic hypothesis, which, of course, does not exclude the influence of environment or its interaction with genetic factors.

Digressing for a moment, Jensen's statement, at the time, was not well received, nor would it receive rave reviews today. Those that disagree with this statement do not necessarily take that position merely because they have their own biases, a "bleeding heart" as some would claim. Here, we show that the evidence simply does not support Jensen's claim.

By way of background, heritability is the proportion of variation in a phenotype (trait, characteristic, or physical feature) that is thought to be caused by genetic variation among individuals. More crisply stated, it is the *proportion of phenotypic variance attributable to genetic variance.* Regardless whether heritability of intelligence is 80% as Jensen claims, 50% as many researchers claim, or much less as yet others claim, the IQ gap between the advantaged and disadvantaged, or, between blacks and whites may be entirely caused by environmental differences. Heritability of individuals may have nothing to do with heritability of groups of individuals.

Moving right along, Flynn reported in 1973, "Jensen states his conclusions about the IQ gap [between blacks and whites] as follows:

In view of all the most relevant evidence which I [Jensen] have examined, the most tenable hypothesis, in my [Jensen's] judgment, is that genetic, as well as environmental, differences are involved in the average disparity between American Negroes and whites in intelligence and educability, as here defined. All the major facts would seem to be comprehended quite well by the hypothesis that something between one-half and three-fourths of the average IQ difference ... is attributable to genetic factors, and the remainder to environmental factors and their interaction with the genetic differences. [Jensen continues,] ...

this means that approximately 8 to 11 points of the gap are genetic and would remain even if the environments of black and white were rendered equivalent."[59] (p. 25)

This statement caused an upheaval when it was published, and rightly so. This chapter is partly written to refute this set of assertions, and that of many other researchers that have followed Jensen. Interestingly, since the writing of those words the black/white gap in IQ has narrowed to about ten or eleven points as documented by Dickson and Flynn in 2006. Accordingly, Jensen's comments would suggest, unreasonably, that, currently, there are hardly any remaining environmental differences causing IQ differences between blacks and whites.

Flynn chooses not to debate the merits of Jensen's position that most of the white/black IQ gap is genetically determined. Rather Flynn focuses on the arguments that intelligence can be modified, and much of the gap can be diminished with these modifications. This seems to me to be an unwarranted and unnecessary compromise on this issue.

Flynn writes,

> The question that is really central is the magnitude of the difference: a genetic gap of say ten points allows a racist to make a last stand in defense of the epistemological viability of his ideology; while if white and black differ genotypically by five points or less, setting aside whether such a difference favors white or black, the racist ideologue has no real foundation for his defense. Five points is the gap which consistently favour singletons over twins, but no one has in recent years become alarmed over that. (pp. 72-73)

In fact, the difference between singletons and twins is explained by the shared and thus diminished environment experienced both prenatally and perinatally by twins. That five-point gap is generally believed to be based on environmental differences—not genetic differences.

Flynn writes, assuming heritability is the issue,

[59] Flynn, p.25 source Jensen, A. R. 1973. *Educability and Group Differences*, New York. Harper & Row. p. 363

If anyone is interested in my own 'best guess' about heritability of IQ, I [Flynn] suspect that we may someday agree
on values between .45 and .55 for white Americans and between .40 and .50 for black Americans. [And Flynn writes
that to highlight the] possibility of a value between .40
and .60, the target for someone who wants to weaken
Jensen's steel chain of ideas. (p. 158)

Flynn reports,

Thanks to high [heritability] estimates, Jensen believes that
genetic factors are far more important than environmental factors within both the black and white populations of
America today and he believes that this renders a genetic
hypothesis about the IQ gap between the races probable.
He [Jensen] does not believe that [heritability] estimates
alone can decide the issue of genetic versus environmental
hypothesis. However, he [Jensen] argues that the proba-
bility of a genetic hypothesis will be much enhanced if, in
addition to evidencing high [heritability] estimates, we will
find we can falsify literally every plausible environmental
hypothesis one by one. He challenges social scientists who
believe in an environmental explanation of the IQ gap between the races to bring their hypotheses forward. [Then
Flynn writes,] Given his [Jensen's] competence and the
present state of the social sciences, the result is something
of a massacre. (p. 40)

This author accepts this challenge. Furthermore, contrary to other researchers, it is asserted that compromise is not in order. Views may be com-
promised when evidence is vague, but when evidence is clear, compromise
only delays resolution. This author's approach in addressing this persistent
and stubborn question is that while genetics surely partly determines intelligence and differences in intelligence among individual humans, *there is,
yet again, no evidence to support the conclusion that genetic factors determine
any part of the black/white IQ gap.* Quite to the contrary, there is abundant

132

evidence, as is shown in the next few pages, to conclude that the black/white IQ gap is entirely caused by limited nutrition, poor health, less than effective learning experiences, low expectations, motivations, and other environmental concerns disproportionately experienced by the disadvantaged.

Immediately, specific data is conveyed that blacks score relatively low on IQ tests because of several quantitatively explicit reasons such as limited nutrition, health, experiences and other environmental considerations. More specifically, these considerations include the black's (the disadvantaged in general) disproportional frequency of LBW babies, their infrequency of prenatal medical care, the limited nutrition of disadvantaged pregnant women, the significantly limited use of breast-feeding among black mothers (the disadvantaged in general), the limited nutrition for many disadvantaged children, the resulting less than effective formal and informal learning experiences, limited education during summer months for disadvantaged children, and more. These observations are not merely asserted but demonstrated with rationally derived values which convey the importance of each consideration. We will, then, convey the many demonstrated ways, environmental interventions, to raise the intelligence of the disadvantaged, which, incidentally, also raise the intelligence of others not disadvantaged, though to a lesser degree.

One last introductory comment— with each environmental remediation, we point out three observations: the enhancement of IQ in the black population (the disadvantaged population), the enhancement of IQ in the white population (the advantaged population), and the resulting decrease in the IQ gap between them. This points out that these interventions appear not oriented to blacks or whites but to the disadvantaged and advantaged. As will become increasingly clear, the IQ gap is not a racial issue but an issue of economic disadvantage.

So, let's now examine the effects of several environmental factors known to affect the differences in intelligence between groups of people, including blacks and whites as well as the disadvantaged and advantaged. Referencing juried and replicated studies, this chapter shows that because blacks are significantly more disadvantaged than whites, and because environmental constraints may be significantly reduced with specific interventions, the IQ gap between blacks and whites and disadvantaged and advantaged may be eliminated. Let's look at the facts.

The observation that inadequate nutrition, poor health, and therefore limits on effective formal and informal learning experiences inhibit intelligent behavior in the moment and the age-appropriate growth of intelligent behavior needs to be broadened. It seems safe to assert that previous comments probably apply similarly across all cultures, races, and nationalities. Fairly straightforward interventions, some mentioned in the previous chapter, have been studied that reduce many environmental constraints. If those interventions were assured to disadvantaged children, how would the distribution of intelligence in the population change? We can begin to answer this question by calculating the impact of these constraints associated with economic disadvantage on the current population, and by projecting the difference expected from intervention. To that end, we re-examine a much-studied disparity.

The distinction between age-independent IQ scores and age-dependent intelligent behavior is fundamental to understanding the gradual formation of the black–white IQ gap in the United States. (These observations may be expanded to apply to all groups of people.) Stated in terms of IQ, the questions of interest here are: Why is the average IQ of blacks 4.5 points lower than that of whites by age four? And, why does the average IQ of blacks gradually and steadily drop from age four to age twenty-four, while the average IQ of whites remains essentially stable? Stated in terms of intelligent behavior, the questions are more concrete: Why, on average, does the intelligent behavior of blacks increase less age-appropriately than the intelligent behavior of whites during the developmental stages through age twenty-four? Additionally, while we know the data shows that the average IQ of blacks is 4.5 points lower than that of whites by age four, is there evidence that explains the gap at age four and that it results from specific causes that contribute to the gradual growth in the gap starting even before birth?

While the data, here, focuses on the IQ gap between blacks and whites, it also explains, more generally, the IQ gap between the advantaged and disadvantaged.

IQ scores of African Americans (blacks) have averaged lower than those of European Americans (whites) since early in the last century. Available evidence indicates that the gap between the two groups' mean IQ scores remained constant until the early 1970s. The average scores differed by about one standard deviation, about fifteen IQ points.

Recently, this fifteen-point gap has been decreasing. Dickens and Flynn report in 2006[60] that "Blacks have gained 4 to 7 IQ points on non-Hispanic Whites between 1972 and 2002. Gains have been fairly uniform across the entire range of black cognitive ability." (p. 913) Over that thirty-year period, for blacks under twenty-five, the "gains on Whites over time did not vary with age but were steady at 5.5 points at all ages." (p. 916). Nisbett affirms the gap had narrowed to about ten points:

> The best evidence we have indicates… that the Black–White IQ gap has lessened considerably in recent decades (Grissmer, 1994; Grissmer, Flanagan, & Williamson, 1998; Grissmer, Williamson, Kirby, & Berends, 1998; Hedges & Nowell, 1998; Nisbett, 1995, 1998). … The gap is… approximately 10 IQ points. (Nisbett, 2005, pp. 302, 303; See also Nisbett, 2009)

Rushton and Jensen (2006, 2010) disagree.[61]

Some suggest that the reduction in the IQ gap between blacks and whites is due to training for the test, cultural changes, changes in dialect, reduction in LBW babies, providing mother's milk more frequently, better nutrition, and other reasons yet to be discussed. Regardless of the amount of the decrease, or the cause of decrease, in recent years, a significant difference in average group IQ remains.

One reason an explanation for the black–white IQ gap has proven so elusive is the common practice of viewing and analyzing the gap as a single number, as an average across all age levels. This averaging conveys the

---

[60] Reference 2006a
[61] For the debate about the decreasing gap, see, for example, Rushton and Jensen (2006, 2010), Dickens and Flynn (2006b), Nisbett (2009), and Sternberg (2005).

misleading impression that the average IQ of blacks falls below that of whites at all age levels by the same amount—by about fifteen IQ points fifty years ago, by perhaps about ten points more recently.

The impression of a fixed gap between the two groups at all ages is false and has severely delayed understanding for generations. Reviews of the evidence[62] show that the IQ gap between blacks and whites gradually and steadily widens with the ages being compared. Specifically, during much of the last century, the average black–white IQ gap was about ten points for four-year-olds and increased with age to about twenty points for twenty-four-year-olds. This pattern was reported as an average difference across all age groups of about fifteen points $(10 + 20)/2$. More recently, the gap appears smaller at each age level. The gap is now known to reach about 4.5 points by age four; it gradually and steadily widens to about 15.5 points by the early twenties, and is reflected, by some, as an overall average difference of about ten points[63] $(4.5 + 15.5)/2$.

Here, the pattern to be explained is not an average gap in IQ scores between blacks and whites across all age levels, but rather two related phenomena: a gap of 4.5 IQ points reached by age four, and the consistent and steady increase in that gap of about 0.6 points each year from age five to age twenty-four, as shown on the right side of Figure 5. Viewing the problem(s) by examining these separable observations yields a much greater potential to understand the causes.[64]

---

[62] Dickens and Flynn, 2006a; Flynn, 2008, pp. 108-111; see also Jensen, 1998, p. 359
[63] Dickens and Flynn, (2006a); Nisbett, (2005, 2009)
[64] It is shown, shortly, that the gap increases gradually through the prenatal and perinatal developmental stages to reach 4.5 points by age four. This has been difficult to show definitively, because intelligence tests for the youngest children take a different form from those used for children older than age four and comparisons of test results for very young children with test results of children age four and older are less than reliable. However, we know that environmental differences starting with the prenatal developmental stage through the stages to age four affect intelligence.

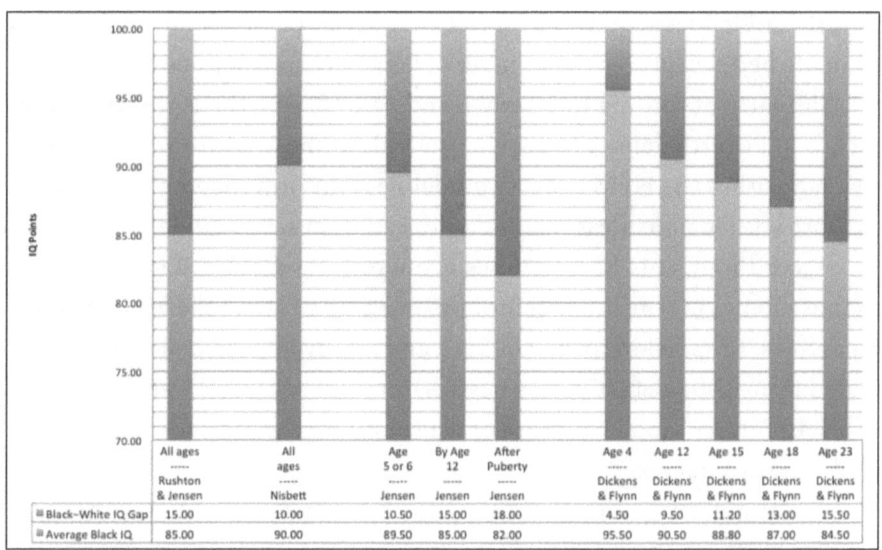

	All ages	All ages	Age 5 or 6	By Age 12	After Puberty		Age 4	Age 12	Age 15	Age 18	Age 23
	-----	-----	-----	-----	-----		-----	-----	-----	-----	-----
	Rushton & Jensen	Nisbett	Jensen	Jensen	Jensen		Dickens & Flynn	Dickens & Flynn	Dickens & Flynn	Dickens & Flynn	Dickens & Flynn
▥ Black–White IQ Gap	15.00	10.00	10.50	15.00	18.00		4.50	9.50	11.20	13.00	15.50
▥ Average Black IQ	85.00	90.00	89.50	85.00	82.00		95.50	90.50	88.80	87.00	84.50

Average non-Hispanic White IQ is set at 100 in all cases, and a standard deviation of the white scores is set at 15 IQ points. Sources: "Thirty years of research on race differences in cognitive ability" by J. P. Rushton and A. R. Jensen, 2005, Psychology, Public Policy, and Law (11), p. 235. "Heredity, environment, and race differences in IQ: A commentary on Rushton and Jensen" by R. E. Nisbett, 2005, Psychology, Public Policy and Law (11), p. 303. The g Factor: The Science of Mental Ability, by A. R. Jensen, 1998, p. 359. "Black Americans reduce the racial IQ gap: Evidence from standardization samples" by W. T. Dickens and J. R. Flynn, 2006, Psychological Science 17(10), p. 916. and "Common Ground and Differences" by W. T. Dickens and J. R. Flynn, 2006, Psychological Science 17(10), p. 924.

**Figure 5. Recent estimates of the black-white IQ Gap in the
United States and of how the Gap increases with age**

Economic disadvantage affects different proportions of the white and black populations

The Health and Human Services Department (HHS) establishes US poverty guidelines based on household income. The guidelines account for family size and are annually adjusted for inflation, but do not consider regional differences in the cost of living. In 2008, the household incomes of about 25 percent of blacks and 8 percent of whites fell below the poverty guidelines[65].

Government agencies use percentage multiples of the HHS poverty guidelines to determine eligibility for need-related services. For example, children in households with incomes of up to 1.85 times the limit set by the

---

[65] DeNavas-Walt, Proctor, Smith, and US Census Bureau, (2010), p. 15, data source: U.S. Census Bureau and Bureau of Labor Statistics, n.d

poverty guidelines are eligible for free or reduced-price school meals[66]. This estimate better captures the number of families at risk for poor nutrition[67].

Similarly, according to Cauthen and Fass (2008), the National Center for Child Poverty defines "poor" households as those with annual incomes that fall within the HHS poverty guidelines and "low income" households as those with incomes less than twice the amount set by guidelines[68].

Accordingly, this book defines economically disadvantaged children as those living in families with low incomes, up to twice the HHS poverty guidelines. Proportions of advantaged and disadvantaged children are estimated from 2008 data. On this basis, 41 percent of all children, 61 percent of black children, and 27 percent of white children are considered economically disadvantaged.

Some might argue that doubling the poverty threshold overestimates the disadvantaged population. However, if we were to define "economically disadvantaged" more narrowly as families with incomes falling within the poverty thresholds—25 percent of blacks and 8 percent of whites—then the average improvements in IQ associated with the provision of mineral and vitamin supplements, breast-feeding, and formal and informal learning experiences to economically disadvantaged children would be considerably higher, because needs in these areas of more impoverished children are more pronounced. The orientation here is that the black–white IQ gap results primarily (but not exclusively) from deficits found in the economically disadvantaged population, both black and white. The economically disadvantaged are likely to be less than adequately nourished and receive less

---

[66] "According to research, food pantries and similar organizations' definitions of poverty are around 180 percent of the poverty line" Chambers and Paynter, (2012).

[67] Fisher, (1992a, 1992b); U.S. Department of Agriculture, (2010); U.S. Department of Health and Human Services, (2011)

[68] In 2008, 19 percent of all children in the US lived in poor households and 41% were in low income families. By ethnic group, 27 percent of white children and 61 percent of black children lived in low-income households, as did 62 percent of Hispanic children, 31 percent of Asian children, and 57 percent of American Indian children. The percentages remained relatively steady over a decade—in the period 1998 to 2008, the maximum, minimum, and mode percentages of children living in low income families were 41 percent, 37 percent, and 39 percent for the entire population, 28 percent, 25 percent, and 26 percent for whites, 68 percent, 57 percent, and 61 percent for blacks. (Chau, (2009), data source: U.S. Census Bureau and Bureau of Labor Statistics, n.d.; cf. DeNavas-Walt et al., (2010); Wagmiller and Adelman, (2009))

than adequate health care. In addition, their formal and informal learning experiences are inhibited by limited well-being as well as different cultural backgrounds[69].

In general, for the purposes of this discussion, black children and white children from households with incomes up to twice the poverty threshold are considered at risk for each constraint on the development of intelligent behavior associated with economic disadvantage. However, where more specific data is available regarding the distribution of a particular environmental factor, the factor-specific risk data is taken into account. This is the case for breast-feeding, low birth weight, and multiple births.

The next section discusses what causes the IQ gap between blacks and whites to form by age four. The following section discusses why the gap gradually and steadily increases to age twenty-four. The discussion focuses on several environmental factors as they relate, respectively, to IQ as measured at age four and to IQ as measured between age four and age eighteen. Existing research quantifying the negative effects of specific environmental conditions on intelligent behavior and its age-appropriate growth is reviewed. Attention is paid to any disproportionate exposure related to economic disadvantage or to ethnic group. An effort is made to determine the extent to which these environmental influences are independent and cumulative, and to assess the extent to which they explain age-related IQ gaps between blacks and whites.

After these data are presented, some questions they raise are considered.

Note, please, that no mention has been made of possible genetic differences between blacks and whites because, to my knowledge, no evidence has ever been presented of genetic *cognitive* differences between blacks and whites or any other racial groups of human beings. Note, also, that Down syndrome and other such genetic issues are prevalent in all races in the same proportions, and that they are prevalent among the advantaged and disadvantaged in the same proportions as well. These are genetic abnormalities, not genetic differences.

---

[69] Moore, (1986); Hart and Risley, (1975); Brice Heath, (2010)

The black-white IQ gap to age four

Average IQ at age four is 4.5 points lower for black children than for white children partly because the incidence of LBW babies among blacks is twice that of whites, and because nutrition among black women is much more inadequate than among white women during pregnancy. Additionally, among blacks, baby formula is provided more frequently than among whites, and also early childhood health and nutrition deficits associated with economic disadvantage are more frequent among blacks than whites; blacks are three times more likely to be disadvantaged than whites. These issues are addressed immediately. Also addressed is the question, how might the black–white IQ gap at age four be affected by specific interventions? It is worth emphasizing, here, that these characteristics are similar among the disadvantaged, regardless of race.

Nutrition during pregnancy

Recognizing that blacks are about three times as likely to be disadvantaged as whites in the United States and given that when nutritional supplements are provided to economically disadvantaged women during pregnancy an 8-point increase in the offspring's IQ is to be expected[70], and providing mineral and vitamin supplements to the disadvantaged, the black–white IQ gap at age four[71] would be significantly affected.

---

[70] As shown by Harrell, R. F., Woodyard, E., and Gates, A. J. in 1955, mentioned earlier.

[71] To estimate the effect of making nutritional supplements available to every low income expectant women, consider that some of the 61 percent of black pregnant women and some of the 27 percent of white pregnant women who are economically disadvantaged already receive prenatal medical care. (Some of them first receive it in the third trimester, and only some receive mineral and vitamin supplements.) Sullivan, Ford, Azrak, and Mokdad reported, in 2009, that 66.4 percent of all black pregnant women and 87.6 percent of all white pregnant women use vitamin and mineral supplements. They convey that usage of these supplemental nutrients is 42.7 percent for the most disadvantaged and generally increases as income increases. They also report 57 percent of those never married use these nutritional supplements. Because the definitive specifics are not known, we assume, perhaps too conservatively, that in low-income families, 60 percent of black pregnant women and 70 percent of white pregnant women currently use nutritional supplements. Providing nutritional supplements to the remaining 40

Please note that much of the analysis for findings is contained in the footnotes to permit the reader to capture the essence of the results and conclusions uninterrupted. Of course, the reader may choose to refer to the analysis in the footnotes at any time.

As seen from the footnote, the average increase in IQ among blacks due to providing mineral and vitamin supplements to pregnant disadvantaged women would be 1.95[72] IQ points. The average increase in IQ among whites would be 0.65[73] IQ points. The reduction in the black–white IQ gap would be the difference (1.95 – 0.65), or 1.3 IQ points. This explains over one-fourth of the current IQ gap at age four. Worth emphasizing, while this intervention would decrease the gap, it would also raise the IQ of both groups. This is shown graphically in *figure 6*.

Incidentally, according to several researchers[74], changes in use of mineral and vitamin supplements among the economically disadvantaged in the last decades of the 20th century may explain part of the drop in the average black–white IQ gap from fifteen to about ten points during those years. If the use of mineral and vitamin supplements among low-income pregnant blacks and whites increased from nil in 1972 to the assumed 60 percent and 70 percent, respectively, in 2002 the average IQ of all black infants should have increased by 1.4 points[75] more than the average IQ of all white infants, simply because of the disproportionate increase in nutrient supplements among blacks. This single factor accounts for about 30 percent of the

---

percent of low-income black expectant women and the remaining 30 percent of low-income white expectant women will increase the average IQs of each group's offspring. Because a larger percentage of the low-income expectant women not already receiving supplements are black rather than are white, extending supplements to all low income pregnant women will benefit a larger proportion of the black offspring and decrease the IQ gap considerably.

[72] 40% x 61% x 8

[73] 30% x 27% x 8

[74] (Sullivan et al., 2009, p. 124; cf. Briefel and Johnson, 2004, pp. 423–424)

[75] Effect on the IQ gap reduction in the most recent generation due to the use of mineral and vitamin supplement by disadvantaged pregnant women, assuming zero usage as the starting point:
Increase in average IQ of offspring among blacks is: 60% x 61% x 8 = 2.9.
Increase in average IQ of offspring among whites is: 70% x 27% x 8 = 1.5.
This would increase average IQ by 1.4 points more among blacks than among whites.

five-point IQ reduction in the black–white IQ gap from 1972-2002, as shown by the research of Dickens and Flynn.

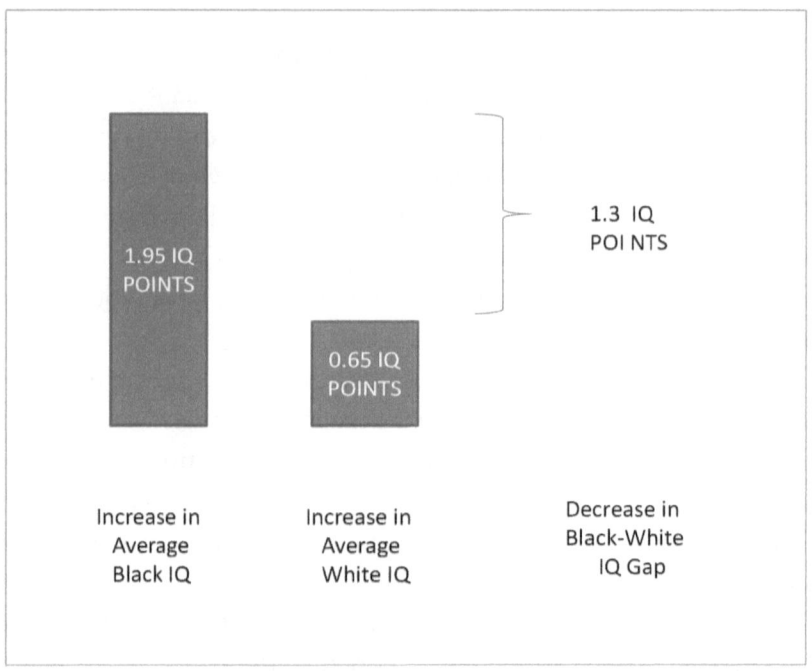

Providing mineral and vitamin supplements to a disadvantaged pregnant woman prevents an 8-point drop in the IQ of her offspring. 46% of Blacks and 15% of Whites are disadvantaged, so this intervention would affect a higher proportion of Black infants. Data from The effect of mothers' diets on the intelligence of offspring: A study of the influence of vitamin supplementation of the diets of pregnant and lactating women on the intelligence of their children by R. F. Harrell, E. Woodyard, and A. Gates, 1955. Bureau of Publications. Teachers College, Columbia University. N. Y., N. Y. Poverty and food insecurity data from Poverty guidelines, research, and measurement by U.S. Department of Health and Human Services, Assistant Secretary for Planning and Evaluation, 2011 and Income eligibility guidelines by U.S. Department of Agriculture, Food and Nutrition Service, 2010.

Figure 6: Expected changes in average IQ of offspring at age four and in the black–white IQ Gap at age four if disadvantaged mothers receive mineral and vitamin supplements during pregnancy

Breast-feeding

Jensen argues in 1998 that the importance of breast-feeding, and of providing LBW babies mother's milk would seem to be highly relevant to the IQ of Black children in contemporary U.S. for two reasons: (1) ... Black infants are much more frequently of LBW [twice as frequent] than are those of other racial/ethnic groups, and (2) they are much less frequently breast fed. ... Black women who breast feed also end nursing sooner than do White mothers. (p. 507)

In recent decades the tendency to breast-feed has increased for both blacks and whites—much more for blacks[76].

If all black and white babies were breast-fed, or, if the same percentage of babies in both cohorts were breast-fed, the IQ gap between black and white children, at age four, would decrease by 1 point[77]. (This assumes that black women and white women breast-feed their babies for the same period of time. Black women breast-feed a shorter average period of time than do white mothers, so if duration were also matched the expected improvement

---

[76] The percentage of babies born in 2005–2006 who were ever breast-fed was 65 percent for non-Hispanic blacks, 79 percent for non-Hispanic whites, and 80 percent for Mexican Americans. Younger mothers are less likely to breast-feed. Among teen mothers, only 30 percent of non-Hispanic blacks, 40 percent of non-Hispanic whites, and 66 percent of Mexican Americans ever breast-fed their infants. Income matters. As reported by McDowell, Wang, and Kennedy-Stephenson in 2008 (pp. 2–4), between 1999 and 2006, only 57 percent of infants born to families with incomes up to 1.85 times the poverty threshold were ever breast-fed, while 74 percent of infants born to families with higher incomes were ever breast-fed.

About 65 percent of black women breast-fed their babies in 2005, up from 25.3 percent in 1987; in contrast, about 79 percent of white women breast-fed their babies in 2005, up from 61.1 percent in 1987. So, while the frequency of breast-feeding has significantly improved for both blacks and whites, a substantial difference remains.

[77] Thirty-five percent of black babies born in 2005 were not breast-fed and did not gain a probable seven points (averaging the 5.8 and 8.2-point increase for males and females, respectively, if breast-fed for six months) of subsequent individual IQ as a consequence. Seven IQ points not gained by 35 percent of black infants yields a 2.5-point (35% x 7) reduction in average IQ over the entire black infant cohort. Of non-Hispanic white babies, 21 percent experienced the same 7-point deficit, resulting in an average deficit for the entire group of white infants of 1.5 IQ points (21% x 7). Missed opportunities for breast-feeding potentially contributed 2.5 – 1.5, or, 1.0 points to the average IQ difference between the two cohorts at age four. Had we used the result reported by Horta and associates of 3.44 points instead of 7 points then the average IQ difference between the two cohorts at age for would have been 0.5.

would be greater. Conservatively, we assume 1 point here.) This is an additional nearly one fourth of the 4.5-point black-white IQ gap at age four. This is shown graphically in Figure 7 as seen below.

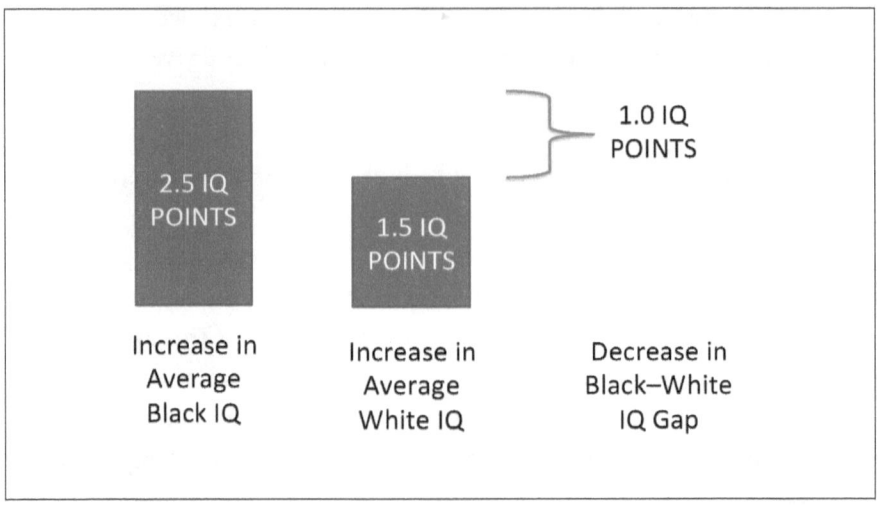

Breast-feeding improves an infant's subsequent IQ by 7 points. As of 2005, 65% of African American mothers breast-fed their children compared with 79% of Non-Hispanic white mothers, so providing mother's milk to all infants not receiving it raises the IQs of 35% of blacks and 21% of whites. Data from "The effect of breastfeeding on child development at 5 years: A cohort study," by P. Quinn, M. O'Callaghan, G. Williams, J. Najman, M. Andersen, and W. Bor, 2001, Journal of Paediatrics and Child Health, 37(5), 465-469 and from Breastfeeding in the United States: Findings from the National Health and Nutrition Examination Surveys 1999-2006 by M. A. McDowell, C. Wang, and J. Kennedy-Stephenson, 2008, CHS Data Briefs No. 5, National Center for Health Statistics, Hyattsville, MD.

Figure 7. Expected changes in average IQ at age four and in the black–white IQ Gap at age four if all children receive mother's milk during the first six months of infancy

First, this chart should not be seen as only addressing the IQ gap between blacks and whites; it is of interest that whites as well as blacks would improve their babies' IQ by increasing the frequency and length of feeding them mother's milk.

There are several other interesting observations to be made about these data. First, there is a great deal of room for improvement. Black mothers can raise the IQ of their children by several points by persisting in their efforts to breast-feed for 6 full months. And the economically disadvantaged, both black and white, breast-feed significantly less than the advantaged.

Many mothers cannot breast-feed for one reason or another. For example, some, especially the disadvantaged, need to return to work and thus cannot breast-feed their babies as frequently. Some do not have enough breast milk. They should know that other means are available to both enhance the well-being and intelligent behavior of their children. And there are various kinds of formula that are differently enhanced.

Changes in the frequency of breast-feeding between 1987 and 2002 may explain an additional part of the recent drop in the average black–white IQ gap from fifteen to about ten points. When the expected gain of 7 IQ points from breast-feeding is multiplied by the percentage increase in breast-fed infants from 1987 to 2002 among blacks and whites, the average IQ of all black infants should have increased by 1.5 points more than the average IQ of all white infants, simply because of the disproportionate increase in breast-feeding among blacks. This one intervention accounts for an additional 30 percent of the overall five-point reduction in the black–white IQ gap during the generation (1972-2002).

The following charts provide additional substantive data.

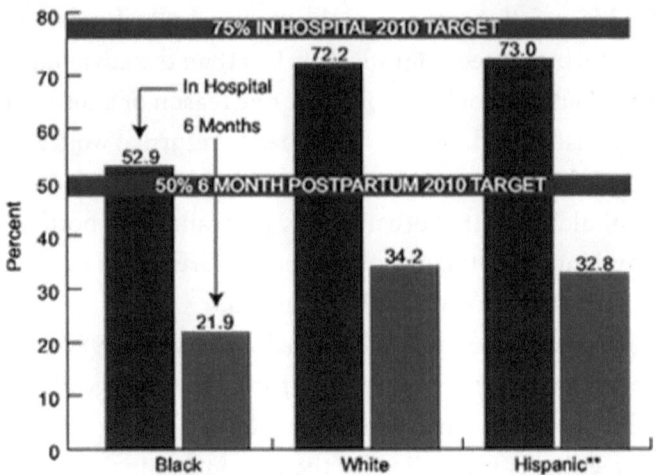

Breastfeeding rates for all women decrease substantially between delivery and 6 months postpartum, the breastfeeding period recommended as most critical for the infant's health by the Surgeon General of the United States. The percentage of women who report that they are still breastfeeding at 6 months postpartum reached a high of 32.5 percent in 2001. At 6 months postpartum, 34.2 percent, 32.8 percent, and 21.9 percent of White, Hispanic, and Black women, respectively, were still breastfeeding.

Figure 8

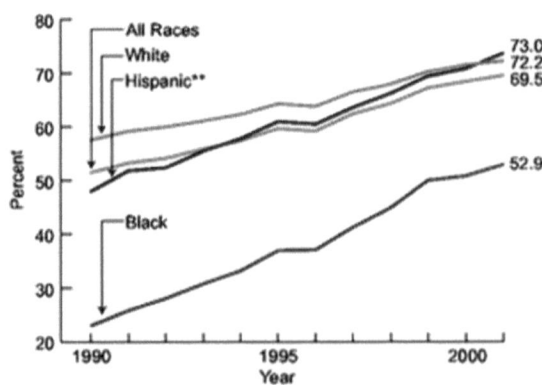

Average breastfeeding rates were highest among women who are over 30 years of age, college educated, and not participating in the Women, Infants, and Children (WIC) dietary supplement program. Overall breastfeeding rates were lowest among women under 20 years of age, Black, low-income, those with less than a high school education, and women living in the southeastern United States.[78]

Figure 9

---

[78] US Department of Human and Health Service, US Health Resources Services Administration, 2003

The effects of increases in breastfeeding and of increased usage of vitamin and mineral supplements during pregnancy help explain nearly three fifths of the reduction in the IQ gap at age four from 1972-2002.

Note that some studies show different data. We used among the most conservative.

Weight at birth

As reported by the North Carolina State Center for Health Statistics and the National Center for Health Statistics, LBW births are about twice as frequent among blacks as among whites, and the disparity in the frequency of very low birth weight (VLBW) births is even greater[79]. Specifically, as a percentage of live births, blacks have more LBW babies than do whites in the ratio of 1.9 to 1, and blacks have relatively more VLBW babies than do Whites in the ratio of 2.7 to 1. Furthermore, the frequency of not-so-LBW births, of infants weighing between 2,500 to 3,000 grams, is also higher among blacks than whites, in the ratio of 1.6 to 1.

Recall, these data are revealing, namely, because blacks are disproportionately more disadvantaged than whites in a ratio of 3 to 1, blacks are more likely to have LBW babies in a ratio of about 2 to 1, and VLBW in the ratio of nearly 3 to 1. The more disadvantaged the more likely births are LBW babies and VLBW babies. This suggests a straight forward intervention – provide disadvantaged pregnant women with better nutrition, or, at least mineral and vitamin supplements, not much different from the supplements prescribed to advantaged pregnant mothers by their prenatal care doctors. With better nutrition among disadvantaged pregnant women there will be fewer LBW babies.

About 4 billion babies are born annually in the United States, of which 8 percent, about 330 thousand, are either born premature, as LBW babies, or both. About 110,000 are born to advantaged women and about 220,000 to disadvantaged women. At an average cost of $50,000 per LBW baby birth[80], the cost to society is about $16.5 billion ($50,000 x 330,000) and significantly decreased capability of each LBW baby in maturity. Nutritional

[79] (State of North Carolina, n.d.; Centers for Disease Control, n.d.)
[80] Source; *Preterm Birth: Causes, Consequences, Preventions*. Richard E. Behrman, Adrienne Stith Butler Editors, Wash. DC: National Academies Press.

supplements provided disadvantaged pregnant women, at a cost of pennies a day, would likely significant decrease the number of LBW babies, and thus decrease the cost of $16.5 billion. A small reduction in LBW baby births would be enough not only to provide nutritional supplements to disadvantaged pregnant women but also to support many other interventions described. If 1 out of 20 of the LBW babies born to disadvantaged women, were born as NBW babies, society would save more than $½ billion (.05(220,000)$50,000 = $550 million) each year. That would be enough to provide mineral and vitamin supplements to all the disadvantaged children in school, from pre-kindergarten through high school. This spotlights the notion that many interventions mentioned in this book are not only self-supportive but support other interventions as well.

Jensen wrote in 1998 that the disproportionate numbers of LBW babies among blacks is thought to be significantly due to their disproportionate disadvantages. The collective effect on the average IQ of blacks is considerable, because the percentage of live births in each of these categories is significantly higher for blacks than for whites, recall that the incidence of VLBW is in the ratio of 2.7 to 1, of LBW 1.9 to 1, and not-so-LBW 1.6 to 1.

Since black mothers are disproportionately economically disadvantaged and more likely to be undernourished, to be mothers as teenagers, and, to receive less prenatal medical care, interventions to correct those problems would disproportionately benefit black children. (It is also possible the incidence of LBW and VLBW babies is higher among black women for genetic reasons, but we lack any evidence to show that.)[81]

Given these data and recalling that LBW babies show an average deficit of 13 IQ points, one can calculate the effect birth weight has on the average IQ gap between blacks and whites. The average IQ deficits associated with VLBW, LBW, and not-so-LBW babies add up to 2.845 for blacks and 1.56 for whites[82], a difference of about 1.3 IQ points (2.845 – 1.56).

---

[81] Providing nutritional supplements to pregnant mothers is likely to reduce the frequency of LBW babies. Reducing teenage pregnancy rates would also reduce the frequency of LBW babies: mothers who are still growing absorb nourishment for themselves to the detriment of their babies (Jensen, 1998).

[82] The percentage of LBW births (excluding VLBW) is 10.5 percent among blacks and 6 percent among whites, for an average deficit of 0.105 x 13 = 1.365 IQ points for blacks and 0.06 x 13 = 0.78 IQ points for whites. The VLBW percentage is 3.2 percent for blacks, 1.2 percent for whites, for an average deficit of 0.032 x 23 = 0.736 IQ points

So, *low birth weight-related deficits contribution of 1.3 points to the IQ gap between blacks and whites is more than 25 percent of the gap, 4.5 IQ points, currently reached by age four.*

The economically disadvantaged have more LBW babies than the advantaged in about the same ratios as for blacks and whites. Thus, this consideration affects the IQ gap between the disadvantaged and advantaged similarly as it does blacks and whites.

Also, of interest is that reducing the frequency of LBW babies would increase the IQs of whites, blacks, advantaged, and disadvantaged, and, likely, other cultural and racial groups, albeit in different degrees.

Time between births

Essentially the same percentage of MZ (identical) twins are born to blacks as to whites, but the percentage of DZ (fraternal) twins born to blacks is about twice the white percentage—1.6 percent compared with 0.8 percent. This is reported, by some, as a reason for low average IQs among blacks yet it actually is a trivial matter. The black–white IQ gap effect is 0.044[83] IQ points, hardly worth mentioning.

As Hunt, in 1961, (p. 362) explains, the constraints on the development of intelligent behavior associated with births in close succession among the economically disadvantaged is also noticed in large families but is not observable "among the well-to-do who can afford to have help in the care of their infants during the early months." He suggests that the constraints are due to "lack of the stimulation that comes from young infants having ample adult contact." This may prove to be another factor in the black–white gap in IQ. Blacks have more children in close succession than whites and have fewer resources to help in the care of their infants especially during the early months. In the State of North Carolina, the frequency of births in close succession among blacks is 15.3 percent as compared to 13.4 percent among

---

for blacks and 0.012 x 23 = 0.276 IQ points for whites. The not-so-LBW percentage is 24.8 percent for blacks and 16.8 percent for whites, increasing the average deficit by 0.248 x 3 = 0.744 IQ points for blacks and by 0.168 x 3 = 0.504 IQ points for whites.
[83] The black–white IQ gap effect is (5.5 IQ points x 1.6%) – (5.5 IQ points x 0.8%), or, 0.044 points.

whites. The incremental black-white IQ gap deficit associated with births in close succession is 0.1 IQ points[84], also not very significant.

Effects of interventions on the IQ gap, through age four

In summary, provision of just these early environmental modifications benefiting economically disadvantaged children of all ethnicities is expected to raise the average IQs of blacks and whites in proportion to the percentage of low income children in need in each group. This diminishes the black–white IQ gap by age four by the number of IQ points shown:

Providing mineral and vitamin supplements to disadvantaged pregnant women ....= 1.3 pts
Providing mother's milk to babies (For example, breast-feeding) = 1.0 pts
Reducing the incidence of LBW, VLBW, and Not-so LBW births= 1.3 pts
Increasing time between births..= 0.1 pts

We have seen that providing mineral and vitamin supplements to economically disadvantaged pregnant black and white women reduce the average IQ gap by 1.3 points, and also that providing mother's milk to black babies and white babies reduces the IQ gap between blacks and whites by one point. As previously discussed, and mentioned in the respective studies, these two effects appear to be independent of each other and are cumulative.

LBW deficits disproportionately affect the black population, and, as we have seen, the effects of the various types of LBW births somewhat overlap with effects of nutritional supplements during pregnancy and breast-feeding of all babies. To underestimate rather than overestimate the combined effects, since about 14 percent of black and 7 percent of white babies born in the United States are LBW including VLBW babies, it is reasonable to assume, for the sake of our discussion, that the mineral and vitamin supplements given to low income pregnant women affect only the NBW population, about 86% of black infants and 93% of white infants. Thus, the

---

[84] The deficit associated with births in close succession is 0.153 x 5.5 = 0.842 for blacks, 0.134 x 5.5 = 0.737 for whites. Thus, the contribution of births in close succession to the gap is at least 0.8 – 0.7, or, 0.1 IQ points.

effects of mineral and vitamin supplements of 1.3 points mentioned for this consideration should only be about 1.1 points.[85]

The effects of births in close succession are independent of whether nutritional supplements are provided pregnant women, babies are breast-fed, or the incidence of LBW babies. Births in close succession contribute negligibly to the black-white IQ gap. The effect is cumulative.

These effects, as modified add up to a 3.5-point reduction[86] in the IQ gap between blacks and whites by age four, without considering the effects of formal and informal learning experiences in these early years, which certainly exist, but for which specific evidence is limited.

Figure 10 shows these results graphically. Notably, Dickens and Flynn (2006a, p. 916) indicate that the IQ gap between blacks and whites at age four is 4.5 points. We have shown that more than three-fourths of the gap at age four appears environmentally explained by just these few considerations.

If we now add other likely effects on the gap due to, for example, formal and informal learning experiences as well as cultural and social differences, there is reason to conclude that the IQ gap between blacks and whites at age four results entirely from environmental differences. Furthermore, these differences cause the gap to form gradually between conception and age four, providing more evidence that the black–white IQ gap is not genetically determined but gradually develops during the developmental stages to age 4. This includes, for the most part, the effect of LBW births which results gradually, though prenatally, due to several environmental influences, such as limited medical care, early pregnancy, and inadequate nutrition.

---

[85] 40% x 61% x 8 IQ points x 86% = 1.68 points of improvement in average black IQ, and 30% x 27% x 8 IQ points x 93% = 0.60 points of improvement in average white IQ, for a relative improvement of 1.08 points by blacks.

[86] The numbers shown in the text are rounded. The actual calculation is: 1.076 + 0.980 + 1.285 + 0.105 = 3.446 points or 3.4 points but 3.5 upon rounding each individual value. The first value, the effect of nutritional supplements to disadvantaged pregnant mothers, was reduced to account for overlap with LBW effects.

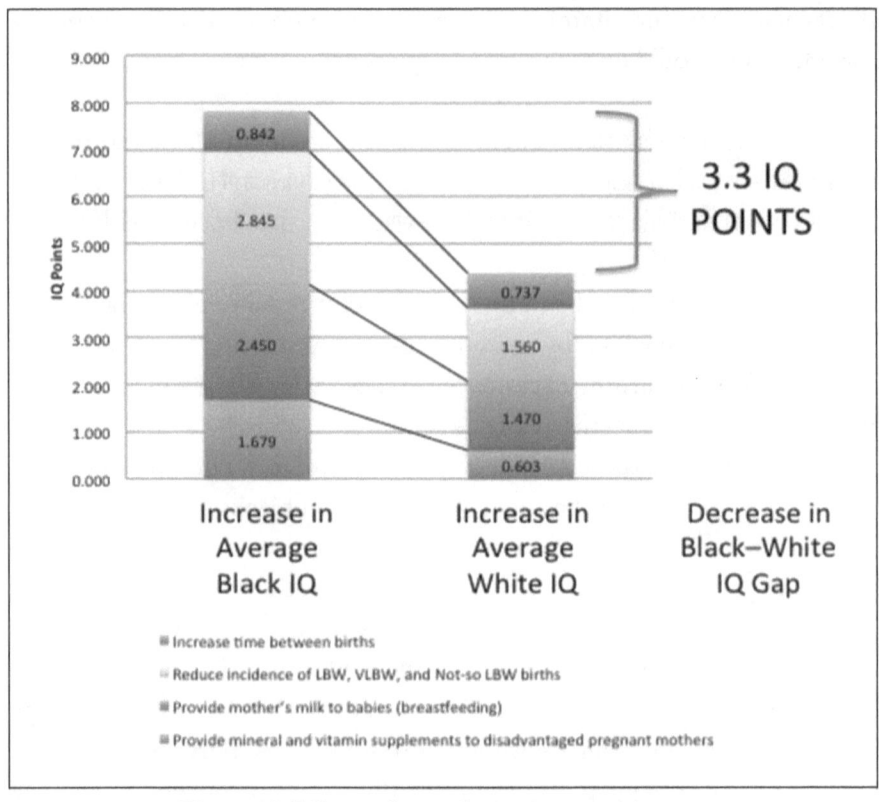

Figure 10. Effects of cumulative interventions to
prevent the black–white IQ gap by age four

Each intervention can be expected to close the IQ gap between black and white four-year
olds by the number of IQ points shown. The effect of supplements to disadvantaged pregnant
women is adjusted downwards to discount that factor for premature deliveries. LBW =
Low Birth Weight. VLBW = Very Low Birth Weight. Sources: The effect of mothers' diets
on the intelligence of offspring: A study of the influence of vitamin supplementation of the
diets of pregnant and lactating women on the intelligence of their children by R. F. Harrell,
E. Woodyard, and A. Gates, 1955. Bureau of Publications. Teachers College, Columbia
University. N. Y., N. Y. "Multivitamin use in pregnant and nonpregnant women: results
from the Behavioral Risk Factor Surveillance System," by K. M. Sullivan, E. S. Ford, M. F.
Azrak, and A. H. Mokdad, 2009, Public Health Reports 124(3), 384–390. Low-income
Children in the United States: National and State Trend Data, 1998–2008, by Chau,
M. 2009, National Center for Children in Poverty, Mailman School of Public Health,
Columbia University, NY, NY. Race, social class, and individual differences in IQ by S.
Scarr. 1981. Lawrence Erlbaum Assoc. Hillsdale, NJ. 2009 North Carolina resident live
births by State of North Carolina, Department of Health and Human Services, N.C. State
Center for Health Statistics, n.d. Retrieved from http://www.schs.state.nc.us/schs/births/
matched/2009/ . Gestation and Birthweight—2009. By Centers for Disease Control,

National Center for Health Statistics, National Vital Statistics System, n.d. Retrieved from http://www.cdc.gov/nchs/data_access/vitalstats/VitalStats_Births.htm. "How much can we boost IQ and scholastic achievement?" by A.R. Jensen, 1969, Harvard Educational Review, 39: pp. 66–67. "Race and IQ: The genetic background" by W.F. Bodmer, 1999 in Race and IQ (Expanded Edition) ed. A. Montagu. 1999, p. 335. Oxford Univ. Press. Oxford, G.B. p. 335. Intelligence and experience by J.M. Hunt, 1961. The Ronald Press Co. NY, NY, p. 362.

Figure 10. Sources and other information

The gradual and steady decrease in IQ scores among blacks after age four is discussed immediately in the following section. Early nutritional and health influences on intelligent behavior establish the foundation upon which subsequent influences build.

The black-white IQ gap from age four to age eighteen

We know that blacks are disproportionately economically disadvantaged and thus likely to be both disproportionately undernourished and less healthy than whites. We also know that the gap in average IQ between blacks and whites steadily widens from age four to age twenty-four, even when children attend the same schools over the same number of years. Is there a causal relationship between these two observations? How might interventions providing adequate nutrition to *all* low-income children while experiencing formal and informal learning reduce the black–white IQ gap from childhood through early adulthood? The same question applies to the IQ gap between the advantaged and disadvantaged. To what extent would just the nutrition intervention reduce the IQ gap between blacks and whites? To what extent would parenting training, richer school break experiences, supplemental nutrition during school breaks, and quality preschool experiences affect these two known IQ gaps? The following addresses these questions. Our focus, first, is on ages four through eighteen.

Parenting education for disadvantaged as well as advantaged parents

As we have seen, economically disadvantaged "parents provide poorer problem-solving strategies for their children than do middle-class parents, and they tend to solve problems for their children rather than assist them in solving them."[87] Moore described, in 1986, similar cultural differences[88] among black and white mothers of adopted black and mixed-race children. Her study consists of two parts. The first part conveys that the black and mixed-race adoptees placed with *black middle*-class families were raised by about fifteen IQ points. This part of the study seems to address the influence of nutrition and health, showing that well-being is enough to bring the IQs of blacks up to that of whites. The second part of the same study indicates that the IQs of the black and mixed-race adoptees placed with *white* middle-class families were raised by about thirty points, that is, an additional fifteen points. Though the evidence does not explain these additional 15 points, this part of the study appears to suggest that cultural influences increase IQs by fifteen points. While this result seems rather strong, it does not detract from the observation that well-being substantially affects intelligent behavior and significantly affects the IQ gap between blacks and whites. Furthermore, these results appear like the Smilansky study described in the previous chapter.

The effects of parenting and culture are tightly intertwined. Based on the range of effects reported earlier from interventions to improve family engagement with children and participation in their education, a gain of ten IQ points for economically disadvantaged children, black or white, is conservatively assumed here. These comments only relate to parenting considerations.

Only the economically disadvantaged are considered here because that is most relevant to determine the effect on the black–white IQ gap. If the 61 percent of black children in low income families and the 27 percent of white children in low income families all benefit, average black IQ will increase by about two times as many points as average white IQ will increase. Providing early parenting education to at least one parent of each economically disadvantaged child is expected to reduce the black–white IQ gap by 3.4 IQ points[89]. Here, again, notably, while the gap decreases, the average IQ of

---

[87] (Ceci, 1990, p. 49, citing McGillicuddy-Delisi, 1982)

[88] Moore (1986) relates the cultural differences she studies to race or ethnicity. Other studies (e.g., Ceci, 1990) associate them with socio-economic status.

[89] Early parenting education to at least one parent of each economically disadvantaged child is expected to reduce the black–white IQ gap by $(0.61 \times 10) - (0.27 \times 10)$ or 3.4 IQ points.

blacks increases by 6.1 points and the average IQ of whites increases by 2.7 points; this due to parenting training.

These estimates only relate to the disadvantaged, but there is abundant reason to claim that the advantaged would significantly benefit from parenting education as well.

Preschool (schooling starting at age four)

The impact of preschool attendance on the IQ gap between blacks and whites is determined by multiplying the expected effect by the percentage of economically disadvantaged black and white children attending preschool. As previously explained, 61 percent of black children and 27 percent of white children are disadvantaged. We do not have adequate data to determine the percentage of these children attending preschool. If no economically disadvantaged children were currently attending preschool, and arrangements were made for all to do so, we would expect the 7-point IQ increase in preschool to yield a reduction in the black–white IQ gap of 2.38 points[90] at the start of kindergarten and 1.02 in the third grade when the IQ increase associated with preschool is only 3 points. Assuming 50 percent current preschool attendance by economically disadvantaged children – both black and white, the black–white IQ gap has already been reduced by 1.19 points at age five and 0.51 points at age eight by providing preschool to those children, and a reduction of the same size can be expected if arrangements are made for all economically disadvantaged children to attend preschool. Conservatively, we use ½ a point to represent the reduction in the black-white IQ gap due to pre-school attendance, at third grade.

Moving along, recall just the observations made by Jensen and Nisbett, if quality preschool were assured its attendees the expected improvements in IQ scores, and more importantly, the improvements in intelligent behavior, would be significantly greater than just mentioned.

The rise in preschool attendance among the economically disadvantaged between 1972 and 2002 is likely to have been yet another of the reasons for the improvement in IQ scores and decrease in the black–white

---

[90] Preschool provides a reduction in the black–white IQ gap by $(0.61 \times 7) - (0.27 \times 7)$ or 2.38 at the start of kindergarten

IQ gap reported by Dickens and Flynn (2006a). Not only Head Start but many other preschool programs yield such improvements. Nisbett, in 2009 (p. 123) reports that a "review of about a dozen of the better small-scale preschool and kindergarten programs focusing on black children showed that they lead to big gains in IQ—of as much as .70 SD [ten points] or even more at age five".

The Milwaukee study, run by Heber and others in 1969, for example, a highly intensive intervention, accounting for a rise in IQ of thirty-five points, may suggest some additional answers. Heber and Garber, the lead researchers of the Milwaukee study, compared three separate groups—an experimental group, a control group, and a sibling group. While the IQs of the four-to-six-year-olds in the experimental group achieved an IQ of a little over 111, these scores dropped to about 105 during the ages of six to ten. This study is often referenced[91]. For instance, Flynn in 1980 (p. 85) indicates, had this group not participated in Heber's program "it is clear that the experimental children would have ended up (at age 14) with a mean IQ of 70 +/− 5 points." Interestingly, some researchers point out that this intervention was costly, but that's true for many studies. The intervention for the children of the well-to-do is also costly, probably no less costly in both time and money.

In 2001, Currie offers an explanation for the drop in reading test scores that may help explain the "fade out" of gains in IQ. (These are verbal scores, which relate to growth of intelligent behavior while being adequately nourished.) In 1995, Currie and Thomas conducted a study of Head Start programs that

> included significant samples of the 60 percent of Head Start children who are not African-American. The estimates of gains for African-American children parallel those of studies in which subjects were randomly assigned, which lends them additional credibility: initial gains in vocabulary and reading test scores "faded out" while the children were still in elementary grades. For white children, in contrast, there were persistent gains in test scores, as well as reductions in grade repetition. It is worth emphasizing that the initial

---

[91] (e.g., Jensen, 1969, 1998; Scarr, 1981; Flynn, 1980)

gains in test scores were the same for whites and blacks—thus, the real difference was not in the initial impact of the Head Start program but in what happened to the children after they left[92].

So, the question suggested here is, what happened after preschool to cause black children to have a different long term IQ gain than the white children?

Thus, while conservative, we use the ½ point estimate as the reduction in the black-white IQ gap. And, please notice, we are only scratching the surface. If the pre-K considerations were viewed in its widest scope including prenatal care, parenting training, and well-being, the numbers mentioned would be considerably higher, but that would confuse the earlier presentations focusing on those interventions.

Reduction in IQ during summer breaks

Conservatively, it is estimated here that, on average, low income children lose about ¼ to ½ IQ points *per* summer break between ages five and eighteen. Over twelve summers this would result in a total decrement of between three and six IQ points for economically disadvantaged children. Assuming, as we have before, that 61 percent of blacks and 27 percent of whites are disadvantaged and likely affected by missed opportunities for growth in intelligent behavior during summers, this explains 1[93] to 2[94] points of the black–white IQ gap. We use the lower estimate of 1 IQ point in what follows.

This may result not from learning experiences directly but rather inadequate nutrition, lacking free or reduced-price meals resulting in less effective both formal and informal learning experiences, but especially informal learning experiences, during school breaks.

Interestingly, a federal program, called *Summer Food Service Program,* has been in place for 50 years but is hardly used. Only 12 percent of eligible children receive meals from this program.

---

[92] (Currie, 2001, pp. 224–225)
[93] The calculation is $(0.61 \times 3) - (0.27 \times 3) = 1.02$ or ~ 1
[94] The calculation is $(0.61 \times 7) - (0.27 \times 7) = 2.38$ or ~ 2

"A significant and substantial decrement in verbal and nonverbal IQ between kindergarten and Grade 12" among poor blacks was first identified as "a fairly linear function of age" and as a "cumulative deficit due to poor environment" by Jensen in 1977 (pp. 190-191). This was, as Flynn (p. 118) describes it in 1980,

> an analysis of sibling data from the total school enrollment (about 1,300 children) of a small rural town in south-eastern Georgia, an area of low socio-economic status. ... The White sample of 653 had a mean IQ of 102 [and, importantly,] showed no decrement in IQ between younger and older siblings, that is, no tendency for IQ to decline between the ages of 6 and 16. The Black sample of 86 [children] had a mean IQ of 71 and did exhibit a decline: over the age range of 6 to 16, IQ of the black children decreased by 1.42 points per year yielding a total decrement of about 15 points.

These decreasing IQ scores likely represent the absence of an expected age-appropriate increase in intelligent behavior, the result of missed chances to acquire the knowledge required to maintain IQ scores, chances lost because children, perhaps, did not attend school or attended school poorly nourished. Adequate nutrition and school attendance have interdependent effects: in combination, they promote the growth of intelligent behavior and the maintenance of IQ. If either is lacking, the effect of the other is much diminished. If an individual does not remain in school or if he or she does not maintain adequate nourishment while in school, then growth of intelligent behavior is inhibited and IQ falls.

Similarly, the gradual increase in the black–white IQ gap between ages five and eighteen can be explained directly by the limited accumulation, by economically disadvantaged children, of the knowledge and skills upon which the IQ tests are partially based (for example, math and language). That limited increase in intelligent behavior is likely indirectly due to limited nutrition among disadvantaged children in general and, thus, more common among the population of black children because their families have dispro-portionately low income.

Data are unavailable upon which to base definitive estimates of the yearly IQ loss associated with the effects of poor nourishment over the thir-teen years of schooling from kindergarten through high school. Consider the result if the annual rate of loss is conservatively estimated at 1.25 IQ points, which is below the 1.42 points lost per year by disadvantaged students while attending school, reported by Jensen in 1977, and of 1.8 points lost per year after dropping out of high school[95]. Using the same methods used above for other environmental factors, over 13 years the average decline among blacks is 9.913 IQ points while among whites it is 4.4 IQ points[96]. In contrast, with adequate nutrition and therefore no decline, the gap in average IQ would decrease by the difference, or about 5.5 IQ points; while the gap drops by 5.5 IQ points, with adequate nutrition the blacks show an effective increase of 9.9 IQ points and the whites show an effective increase of 4.4 points.

The current free school breakfast and lunch programs provide a foun-dation for school success for less than well-nourished economically disad-vantaged children. While these programs help significantly, they do not resolve the problem. There is reason to believe that these programs require improvement. For example, arguably[97] 85 percent of blacks are lactose in-tolerant as compared to about 15 percent of whites. It seems likely that the affected children avoid consuming the milk provided, which incidentally accounts for a large part of the protein content in school meals. And, do we know what the children really eat? Should soy milk be offered instead of milk? Do these two meals, breakfast and lunch, provide enough nutrition? What about nutrition outside of the school day? (Efforts by a recent first Lady, Michelle Obama, addressed some of these questions.)

The average IQ of blacks gradually decreases throughout youth. Missing

---

[95] Härnqvist, (1968a, 1968b)

[96] As before, we define the disadvantaged population at risk of poor nourishment to be 61 percent of blacks and 27 percent of whites. The average *annual* rates of loss would be 1.25 x 0.61 = 0.763 points in average black IQ lost and 1.25 x 0.27 = 0.338 points in average white IQ lost. Over 13 years, 1.25 IQ points lost annually by 61% of Blacks yields a total decline in the average black IQ of 1.25 x 13 x 0.61 = 9.913 IQ points. If 27% of whites lost IQ at the same rate over 13 years, average white IQ would decline 1.25 x 13 x 0.27 = 4.388 IQ points.

[97] This estimate of lactose is under debate

school—either by not attending school or by attending school with less than adequate well-being—results in a significant gradual loss of IQ. The evidence strongly suggests that, on average, black children are not getting as much out of their school years as are white children, and that a gradual decrease in IQ scores results, directly and indirectly, primarily from lack of adequate nutrition and health even while attending school. (Some loss is perhaps also due to differences in motivation and due to differences in expectations by both teachers and parents. That part relating to parents may be best achieved through parenting training programs. Teacher training needs to address motivation as well as expectations.)

A story about one of the children I tutored and mentored is relevant here.

> I had been working with Phil for about two years. He was labeled learning-disabled and was extremely shy. He would walk down a school hallway staring at the corner where the wall met the floor so that he would not see anyone and would not have to say hello. When I met him at age 15, he could not read and saw no reason to change that. Gradually he caught up with academic demands, but he remained excruciatingly shy. One important event was life changing in many ways, and especially with regard his shyness. Phil never asked a question in class, never participated unless absolutely necessary. Now he had to present a paper to the class. He would not, could not, do that. He finally agreed to at least write the paper. After that was done, using his outline, I got him to write a list of important points and sub-points. We then discussed those topics. Gradually he became comfortable presenting them to me. Then he cut the list down to fit the time-slot. He practiced several more times and learned that it still worked even if he forgot or left out a point. Finally, he gave the talk, without notes, to the class. He told me later that as he walked back to his desk he heard one of the students say to another, "He *is* smart." Phil will not forget that feeling. Nor will I.

What's notable, about this story, and other stories mentioned in this book, is that these children, despite being at least somewhat undernourished – their well-being and thus intelligence therefore being likely constrained – are significantly helped by just raising expectations, motivation, and confidence. Phil did not live in poverty; he was not undernourished. But he was labeled learning-disabled early in his childhood. Expectations were encouraged to be low. Yet, by reasonably raising expectations and gradually enhancing self-confidence, his intelligence and capability was significantly increased. While this is true for learning-disabled children, these observations are likewise true for the entire spectrum of children. High expectations, motivation, and confidence drives children as well as adults, the poor as well as the rich, to more closely reach their potential. Though we do not know how to measure the effect of motivation, confidence, and expectations, we do know they influence people, the young and older, to increase their experiences and thus enhance their intelligence.

Effects of interventions from age four to age eighteen

In summary, provision of these formal and informal learning experiences *together* with ensuring that children of all ethnicities from low income families receive regular, adequate nutrition, is expected to foster more effective age-appropriate growth of intelligent behavior among the economically disadvantaged. This would, accordingly, at least reduce if not eliminate the decline in IQ; this will raise the average IQs of blacks and of whites (in proportion to the percentage of economically disadvantaged children in need in each group), and thereby reduce the black–white IQ gap at each age through age eighteen.

The growth of the IQ gap between blacks and whites from age 4 through age 18 can be reduced or prevented through the effects of four interventions on behalf of economically disadvantaged children regardless of ethnicity: early parenting education for primary caretakers addressing an array of considerations including motivation, expectations, sensitivity, and an array of differences in culture associated with differences in class or ethnicity (3.4 points), providing children adequate nutrition while receiving formal and informal learning experiences that allows them to more effectively and more fully benefit from pre-school (0.5 points), educational experiences during summer breaks (1.0 points), and 13 years of K–12 schooling with an underlying adequate well-being (5.5 points). These interventions add up to a nearly 10.5-point decrease in the black–white IQ gap.[98] (See *figure 11*.) These measures, together with the interventions discussed for the prenatal and perinatal developmental stages, could essentially eliminate the black–white IQ gap.

The effectiveness of formal and informal learning experiences depends on a well-nourished and healthy brain. Inhibited growth of intelligent behavior due to limited nutrition and health helps explain the gradual and steady decrease in average IQ scores of the disadvantaged in general. However, when an individual's IQ score decreases, it need not indicate merely a decrease in potential intelligent behavior. The decrease in scores may mark a lack of age-appropriate growth in intelligent behavior which, in time, and with adequate nutrition, may be recovered.

This review of several distinct but related aspects of formal and informal learning experiences suggests several questions. Though the effects appear independent, do those effects cumulatively help explain the IQ gap between blacks and whites? Do they explain the gains from participation in pre-school programs and their reduction from seven to three points by third grade? Do they explain the steady and gradual reduction in average IQ scores of the economically disadvantaged from age four through adolescence?

---

[98] The numbers shown in the text are rounded. The actual calculation is: $3.400 + 0.510 + 1.020 + 5.525 = 10.455$ points or 10.5 points after rounding.

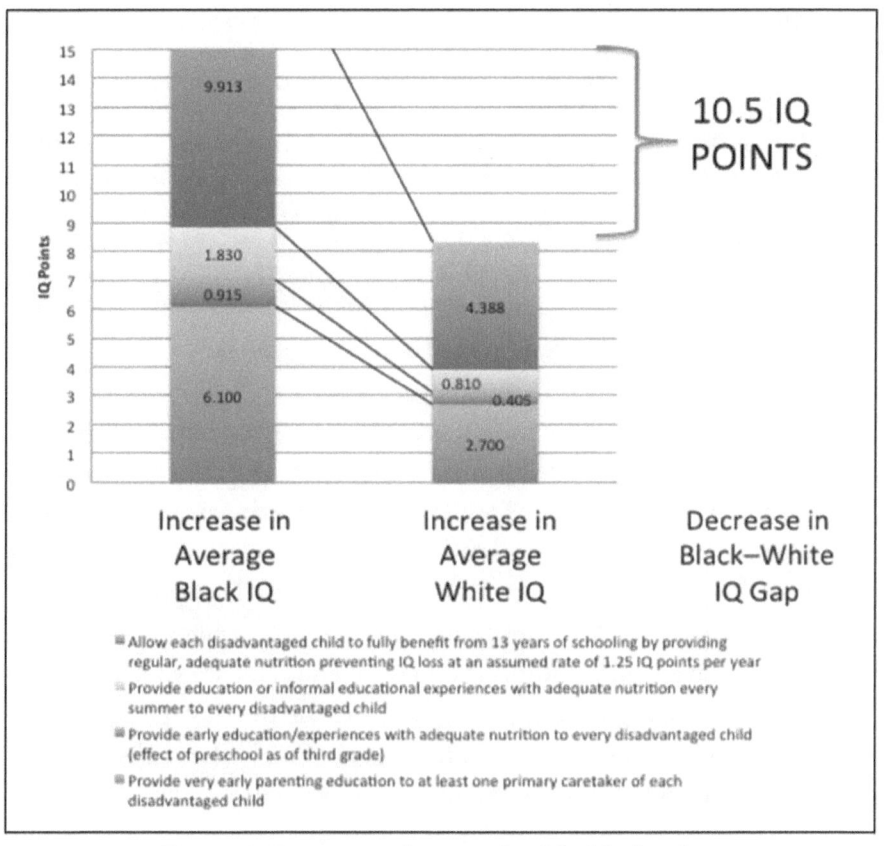

Figure 11. **Preventing the growth of the black–white IQ gap between age five and age eighteen**

Each intervention can be expected to reduce the growth of the Black–White IQ gap between ages 5 and 18 by the number of IQ points shown if provided to all disadvantaged children in need. 61% of Black children and 27% of White children are disadvantaged, so each intervention would affect a higher proportion of Blacks than Whites and disproportionately increase the average IQ of Blacks. Where reports of the effect of an intervention effect vary, a value from the low end of the range of available estimates of effectiveness is used. Providing early education in parenting skills to at least one primary caretaker of each disadvantaged child is assumed to add 10 points to the child's IQ. Preschool attendance increases the IQ of disadvantaged children by 3 points as of the start of third grade. Current preschool attendance among disadvantaged children is assumed to be 50%; the effect of 100% attendance over 50% attendance is shown. Over 12 summers, .25 IQ points are assumed lost per year without formal or informal summer educational activity. Over 13 years, 1.25 IQ points are assumed lost per year of schooling without adequate nutrition. Race, IQ, and Jensen by J. R. Flynn, 1980, London: Routledge & Kegan Paul, pp. 118, 175–181. "Environmental effects on language development: A study of young children in long-stay residential nurseries" by B. Tizard, O. Cooperman, A. Joseph and J. Tizard, 1972, Child Development, 43(2), pp. 337–358. "Is early intervention effective? Some studies of early education in familial and extra-familial settings" by U. Bronfenbrenner, 1999, in A. Montagu (Ed.), Race and IQ. Oxford, UK: Oxford University Press. The bell curve: Intelligence and the class structure in American life by R.J. Herrnstein and C. Murray, 1994, Simon and Schuster. New York, NY, p. 405. "How much does schooling influence general intelligence and its cognitive components? A reassessment of the evidence" by S.J. Ceci, S. J., 1991, Developmental Psychology, 27(5), p. 705. Low-income Children in the United States: National and State Trend Data, 1998–2008, by Chau, M. 2009, National Center for Children in Poverty, Mailman School of Public Health, Columbia University, NY, NY.

Figure 11. **Sources and other information**

A principle and an observation together provide an answer to these questions: *Year by year, if intelligent behavior fails to grow age-appropriately, IQ scores progressively decrease. Both black and white disadvantaged children experience environmental constraints. Black youth are disproportionately exposed to environmental constraints limiting the growth of intelligent behavior.*

Interestingly, while little specific data is available reporting the IQ gap between the advantaged and disadvantaged, enough data shows that the above observations are similarly relevant to explain the IQ gap between the advantaged and the disadvantaged, regardless of race. These observations help improve the IQs of the disadvantaged, in general.

Early positive nutritional and birthing interventions have been shown to cumulatively reduce or prevent the IQ gap between blacks and whites by age four. The resulting IQ gap because of a very few interventions has been shown to be 1 point (4.5 points – 3.5 points). These considerations only deal with the earliest developmental stages, and with the health and nutritional well-being of the brain. These only indirectly reflect early effects on formal and informal learning experiences which, as discussed, are strongly influenced by good nutrition and well-being, in general.

The combined effects of the encouragement and attention that results from early parenting education, and (with adequate nutrition) the educational experiences of preschool, summer programs, and schooling from kindergarten through high school, can prevent the black–white IQ gap from gradually increasing from age five through age eighteen.

Formal and informal learning experiences affect the economically disadvantaged differently from their effect on the advantaged simply because of their different well-being. Blacks, being much more likely to be economically disadvantaged than whites, benefit less, on average, from formal and informal learning experiences—not because of different learning opportunities (though that may remain a concern) but because limited nutrition and health impede learning.

As shown earlier, unlike whites, the IQ of blacks (who are three times as likely to be disadvantaged) decreases at a steady rate of 0.6 points per year after the age of four. Not surprisingly, the data just described explain a steady yearly decline. Note also that some of the data explaining a decrease of 10.5 points include some interventions such as parenting training and pre-K benefits that partly apply to earlier periods – before the age of four.

When 10.5 points for the interventions applied from age five to eighteen are added to the previously-mentioned early environmental modifications amounting to 3.5 points, these interventions alone are collectively enough to just about eliminate the current, or, more accurately stated, prevent the black–white IQ gap at age eighteen. (See *figure 12.*)

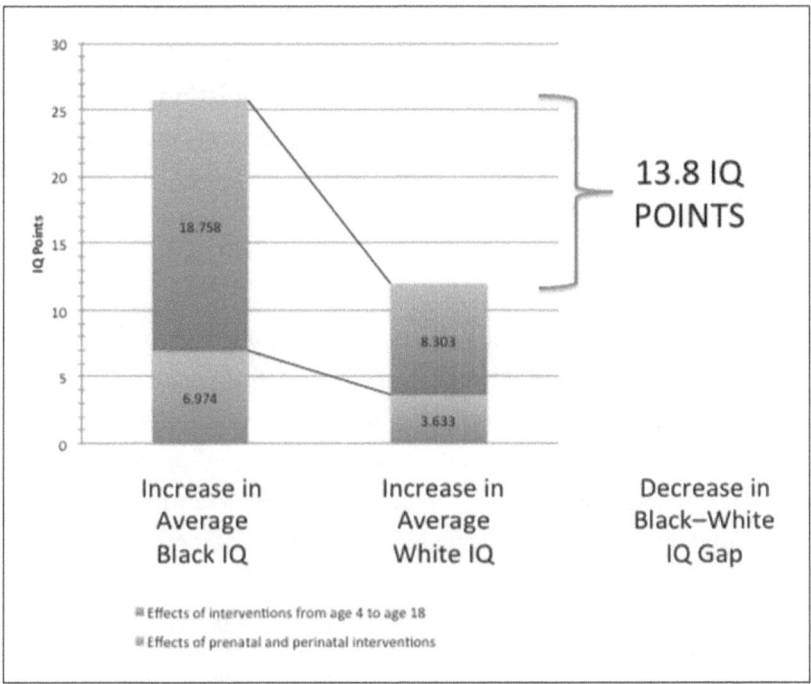

Figure 12. **How the interventions discussed in the text are expected to change average IQ and reduce the black–white IQ Gap by age eighteen**

The effect of these interventions, a reduction in the black–white IQ gap of 14 IQ points by age eighteen, exceeds the 13-point magnitude of the gap at that age reported by Dickens and Flynn[99] yet falls below Jensen's estimate of 18 points reported in 1998 (p. 359). Note, Jensen's estimate was made prior to the report by Dickens and Flynn reducing the gap by about 5 points. The range of these estimates of the gap at age eighteen reflects different assumptions, different data sources, different populations considered, and different error expectations. The relative size of the economically disadvantaged population used for calculations here are also uncertain estimates: families' economic circumstances fluctuate, as do children's exposure to environmental constraints on intelligent behavior and its growth. While several reductions were conservatively estimated, some considerations were incorporated in broader topics. For example, the value of motivation and expectations were subsumed in parenting training. Also, the value of improved learning experiences before age 4, conservatively placed by some at about two points, was not directly included.

Average black IQ continues to drop by about 0.6 points every year from the ages of nineteen to twenty-four, increasing the black–white IQ gap. This is likely due to the fact, as Fry indicated in 2011, that in the years after age eighteen, a much larger proportion of the white population attends college, so that blacks on average receive significantly less of the formal and informal learning experiences on which IQ tests are based. [100]

Other factors

Some considerations that are known to affect the gap were not specifically reflected in the analysis either because of lack of definitive evidence showing their effect on intelligent behavior, or because the evidence showed their effect on the IQ gap to be trivial. For example, the nutritional estimates only include mineral and vitamin supplements—not, for example, the effects of protein supplementation. Also, the effects of prenatal medical care, dental

---

[99] 2006a, p. 916; 2006b, p. 924; cf. Nisbett et al, 2012 and Flynn, 2008, pp. 97–111

[100] Fry (2011, p. 4, cf. p. 13) also reports that "in 2010, 38% of all 18- to 24-year-old blacks were enrolled in college, up from 13% in 1967 and 32% in 2008." This growth in enrollment provides yet another explanation of the reduction of the IQ gap between blacks and whites in the last generation.

care, perinatal medical care, mother's birthing age, and stress unique to the economically disadvantaged have been mentioned but not analyzed because quantitative evidence of their effects on intelligent behavior are limited or unavailable. Several other factors that surely influence individuals have not been directly considered because the size of the effects combined with their statistical distribution in the population, black and white, low income and medium-to-high income, indicates that at best they would be associated with a relatively trivial impact on the gap.

For example, regarding the effects of fetal alcohol syndrome (FAS), in 1998 Abel (pp. 195–201) writes:

> In studies in which the population examined was primarily of middle SES and Caucasian, the incidence rate [of FAS] was 0.26/1000, which is very close to the incidence rate in Europe. By comparison, the incidence rate at sites where the patient population was primarily of low SES, and was African-American, the rate was 10 times higher, 2.29/1000 (Abel, 1995). Since FAS occurs in all racial groups (Abel, 1995), apart from the artifact previously mentioned, race is not a primary risk condition acting in conjunction with alcohol abuse. The most likely alternative is that FAS arises out of a combination of alcohol abuse and poverty[101]

Thus, while surely significant to everyone affected, and much more prevalent among blacks than whites (and more prevalent among the economically disadvantaged), FAS has a trivial effect on the black–white IQ gap, an effect less than 1/50th of the effect of different frequency of blacks and whites having LBW babies.

Similar comments may be made for other environmental factors associated with disadvantage in other countries but not in the United States, for example, limited iodine intake and iron deficiency effect intelligence in several countries but not the United States.

---

[101] (Abel, 1995; Abel and Hannigan, 1995).

Several specific long-term studies

Studies of several other environmental differences are now cited. For example, Scarr and Weinberg reported on a study of disadvantaged children adopted into middle-upper class homes at six months versus twenty-four months of age. The difference in their IQs was 13.5 points.

The Eyferth study is now discussed.

Eyferth study; after World War II

Occasionally studies are made possible through an accident of circumstances. One such study took place in Germany after World War II. This study reported by Herrnstein and Murray, in 1994 (p. 310)

> came about thanks to the allied occupation of Germany following World War II, when about 4,000 illegitimate children of mixed racial origin were born to [white] German women. A German researcher, Eyferth, tracked down 263 children of black servicemen and constructed a comparison group of 83 illegitimate offspring of white occupation troops. The results showed no overall differences in IQ. The actual IQs of the fathers were unknown, and therefore [Herrnstein and Murray concluded] a variety of selection factors cannot be ruled out. The study [they claimed] is inconclusive but certainly consistent with the suggestion the Black and White IQ difference is largely environmental.

In fact, while not statistically significant, Lewontin, Rose, and Kamin, in 1984 (p. 126) reported there was a "small difference favoring the black children."

Here the environmental factor of interest is that all these children, whether fathered by a black or white man, were mothered by white women. This study, thus, supports several related positions that intelligence is substantially affected by environmental factors and, most relevant to the subject at hand, that there is no difference in the genetic influence of intelligence between blacks and whites.

Interestingly, Jensen refutes these findings because there is no knowledge of either the mother's or the father's IQs. More interestingly, Jensen (1998, p. 483) points out another related and interesting statistic, "the white and black fathers were not equally representative of their respective populations, since about 30 percent of blacks, as compared with about 3 percent of whites, failed the pre-induction mental test and were not admitted into the armed services."

Stated differently, in 1975 Loehlin et al. write

> The U. S. black fathers would have been somewhat positively selected for IQ relative to the total U. S. population – the rejection rate for blacks on the pre-induction mental tests during World War II was in the neighborhood of 30 percent compared with about 3 percent for whites, but the difference remaining between average black and white inductee on the Army General Classification Test was at least one standard deviation (Davenport, 1946, Tables I and III). (p. 127)

While these explanations seem, at first glance, to be rational, and even, for some, explains away the study results, they do not hold up to scrutiny.

Specifically, we know that black inductees were selected from a population that had an average IQ of about 85. The 30 percent that were rejected to serve, very likely, had IQs below 85. As to the remaining population, the right most column, as shown in the following table, shows the percentage of blacks in each of three IQ categories, whereas the middle column shows the corresponding percentage of whites.

After taking the pre-induction test, 30% of blacks and 3% of whites were not admitted into armed services; 70% of blacks and 97% of whites were inducted. The following chart shows that the percent of black inductees who had an IQ below 85, and, who had an IQ between 85 and 100, was significantly higher than white inductees. The percent of black inductees who had an IQ above100 was much lower than white inductees.

	Entire Population	White inductee Population	White Study Population	Black inductee Population	Black Study Population
IQ below 85	16%	16–3=13%	13/97 = 13%	50-30 = 20%	20/70= 29%
IQ from 85 to 100	34%	34%	34/97 = 34%	34%	34/70= 49%
IQ above 100	50%	50%	50/97 = 52%	16%	16/70= 23%

Accordingly, *the black fathers of the study sample would likely have had IQs averaging about 93, considerably lower than the white fathers of the study averaging a little above 100.* While Jensen's observation seems valid, it is highly misleading and irrelevant. The children with black fathers in the study would have been expected to have had a lower IQ than the children with white fathers, but they did not. This study provides additional evidence that the differences in IQ between blacks and whites are not genetically but rather environmentally determined.

Interestingly, in 1980 Flynn reports that Eyferth,

> supplied considerable detail concerning the German mothers. They were mainly (although not exclusively) from lower classes, tended to be young, many of them illegitimate themselves, and some have been pushed into relations with soldiers by their own mothers. ... Some of these mothers were themselves in homes for problem youths. The parents of mothers from the working class tended to be more accepting, but most of them were raising their children under difficult conditions. Their incomes were low and most lived in substandard neighborhoods, many in 1960 were still living in barracks, housing for the unemployed or state housing. Many were living with their own parents with much overcrowding; for example, the children almost never had a room of their own. The children also suffered from less tangible handicaps. Eyferth found that Germans had strong feeling against illegitimate children and those

whose mothers had been 'American-lovers' and had had affairs with the enemy. He hypothesizes that the black children may have had their own problems because their skin colour advertised their position so clearly. (pp. 86-87)

Yet the IQ score means of the children with black fathers and white fathers were essentially identical.

Dichotomy of more white boys occupying high IQ levels than white girls whereas more black girls occupy high IQ levels than black boys

I insert here a strange dichotomy. In 1980, Flynn (p. 188) highlights an important study reporting that at high IQ levels, 120 and above, African-American girls outnumber African-American boys by ratios of at least 2 to 1. "Jensen argues that the surplus is merely the result of a small difference between the sexes in terms of their mean IQs ... a small difference at the mean can engender a large difference at the extremes." Interestingly, Flynn (p. 189) continues, Sowell's "evidence on the white population also show the mean for girls above the mean for boys; and yet, at high IQ levels, there are fewer white girls than white boys." What would explain this dichotomy? Why are there twice as many black girls with high IQs than black boys, whereas there are fewer white girls with higher IQ scores than white boys? The answers are not evident. Perhaps the reason is culture, motivation, opportunity, self-confidence, or any combination of the above. Perhaps it relates to caretakers' influence. I suspect a real appreciation of this observation will advance our understanding of the black/white gap in IQ and, thereby, also, our understanding of intelligence in general.

The black-white IQ gap results from disadvantage

We reviewed research on environmental factors associated with reductions in the IQ of economically disadvantaged children and learned that the loss of IQ points, by age four, is largely prevented primarily with few interventions, namely by:

a. Providing low income pregnant women with mineral and vitamin supplements,
b. Encouraging mothers to feed their children mother's milk,
c. Taking steps to limit the incidence of LBW babies,
d. Assuring early parenting education for all high school students, and potential parents,
e. Discouraging births in close succession.
f. Assuring adequate nutrition and good health for all children.
g. Reducing nutritional and health constraints on the growth of intelligent behavior, to age four.

The loss of IQ points between age four and age eighteen can be prevented by

a. Assuring adequate nutrition and health, including mineral and vitamin supplements, so that formal and informal learning experiences are more effective.
b. Providing early parenting education.
c. Extending pre-school education as well as quality pre-school education,
d. Assuring adequate nutrition and health during school breaks.
e. Making formal and informal learning activities available to all children during school breaks.

The black–white IQ gap to age eighteen is largely the result of inadequate nutrition and relatively poor health associated with economic disadvantage. Limited well-being affects both the maintenance of intelligent behavior in the moment and the growth of intelligent behavior over time. By ensuring adequate nutrition while providing formal and informal learning experiences, the effects of environmental constraints can be corrected for both low-income blacks and low-income whites; the IQ gap between these populations can be prevented for coming generations. The IQ gap is also influenced by cultural factors which are partly and indirectly affected by being disadvantaged.

Although modifications to educational systems might be useful for several reasons, this book assumes the current educational systems. Two educational changes are mentioned that affect the black–white IQ gap as well as the IQ deficit associated with economic disadvantage: 1) to provide parenting training, which addresses an array of considerations including some cultural considerations, and 2) to provide availability of school break programs for all. (There is reason to believe that the need for parenting training is not unique to the economically disadvantaged.)

The widening of the black–white IQ gap between age eighteen and age twenty-four appears also to result largely from the disproportionate exposure of low-income young adults to poor nutrition and health during a period of life when formal and informal learning activities continue to contribute to the growth of intelligent behavior, as measured by IQ tests.

The environmental interventions we have discussed would likely entirely prevent the IQ gaps between blacks and whites as well as between the economically advantaged and disadvantaged. These interventions would directly enhance the intelligent behavior of the economically disadvantaged and less directly enhance the intelligent behavior of all by allowing all to be taught with higher expectations[102].

Broad implications of these findings

We have shown that the IQ gap between blacks and whites may be eliminated with just environmental interventions. We have also shown, though subtly, that intelligence may be raised by more than an average of twenty-five points among blacks and by an average of twelve points among whites; this focusing essentially only on raising the IQ of the disadvantaged. Furthermore, because most of the world's population is disadvantaged, the interventions shown in this chapter would result in a greater than twenty-five-point increase in IQ; where the portion of the population that is disadvantaged is 27 percent, as is the case for whites in the United States, the gain in IQ would be an average of twelve points.

These numbers are of profound significance when discussing many of society's problems, some of which being the problems of society discussed in the next chapter. Specifically, we know how to raise the IQ, especially and most substantially, of populations that are significantly disadvantaged. How would that, could that, effect the various conditions of society?

---

[102] Cf. Ceci and Papierno, (2005)

Different researchers seek answers to explain the IQ gap in different ways. Those that assume that the gap between racially different groups of individuals can be entirely explained by environmental differences seek environmental explanations to support their claim. Those that assume that the gap is partly, or mostly, determined by genetic factors seek genetic explanations. This later group examines skull size, immigration from Africa populating the world, and ancestry of blacks to explain possible genetic differences. See Jensen and Rushton's 2005 work for details on these explanations. Of course, smaller skull sizes may result directly from limited nutrition especially during the prenatal and perinatal developmental stages. Other genetic explanations documented by Jensen and Rushton are debunked in an article written in 2005 by Sternberg, then of Yale University.

As stated often enough, surely there are genetic differences among the races (whatever that might mean), and it is possible that there are genetic differences in intelligence among the races. But what if those estimates only conveyed differences—not one being better than another? What if the biological, ancestral, and migration reasons conveyed by some to explain why blacks are genetically different explains only that; that the races are merely different in some ways?

The effects of several clearly identified environmental differences between blacks and whites were examined to understand the extent by which they determine the differences in IQ between blacks and whites. As those *environmental differences have been shown to collectively account for the entire fifteen-point difference in IQ between blacks and whites in 1970, and ten-point average difference in IQ between blacks and whites in 2002, it is concluded, accordingly, that the black/white IQ gap is caused by environmental differences.* It is not relevant that intelligence or IQ of *individual* human beings is partially determined by the environment and partly by genetic factors. What is relevant, and is concluded here, is that *to minimize the difference in IQ or intelligence between blacks and whites, one should minimize the differences in specific environmental factors.*

Chapters 4 conveyed the awareness that the potential intelligence of all of us, but especially of the disadvantaged, can be more closely achieved through environmental enhancements. This chapter conveyed that the black/white IQ gap can be eliminated with just environmental enhancements. This has been shown to relate to the IQ gap between the advantaged and disadvantaged, as well. With current knowledge including the environmental influences and known genetic influences, the message is that the IQ gaps between the groups are not even partly genetically determined, but rather, the black-white IQ gap and the advantaged-disadvantaged IQ gap are entirely environmentally determined and therefore need not persist.

Is there any scientific legitimacy in declaring, as did Herrnstein and Murray in 1994 that a society needs to establish a separate society for those with limited intelligence, so they are kept from underfoot and so that the rest of society can more effectively go about its business of everyday life? There are intense moral questions with their declaration, but more to the point in this discussion, there are at least two critical scientific problems with the statement by Herrnstein and Murray. Namely, their statement asserts, with unjustified near absolute certainty, that the intelligence of people with acknowledged low intelligence is largely, or substantially, genetically determined, and, their statement glibly deals with averages and the associated severe problem of finding the line separating those with low intelligence from those with *adequate* intelligence, whatever that may mean. The problem is so offensive to some that they cannot deal with it rationally. We have tried to approach the problem with unemotional reasoning.

An analogy places these comments in perspective. One's doctor does not say, 50 percent or 80 percent of your problem results from causes about which nothing much can be done. Rather, she or he says that the problem may result from several causes some of which we cannot do anything about. We'll take some tests and find out the causes and then act. In contrast, Herrnstein and Murray, and others assume and claim that because intelligence of individuals is partly determined by genetic factors (they say largely determined by genetic factors), there is not much that can be done, when, in fact, much can be done. Furthermore, and repeating yet again, while genetic factors surely play a role in differences in intelligence between individuals, there is *no* evidence, none whatsoever, that genetic factors play a role in differences in intelligence between groups of people.

It is time to recognize that both the black/white IQ gap and the advantaged/disadvantaged IQ gap are entirely determined by environmental factors. There is no justification to claim genetic factors explain the group differences in intelligence between the disadvantaged and advantaged. And there is no justification to claim genetic factors explain group differences in intelligence between blacks and whites. The scientific community should no longer accept these unfounded claims.

CHAPTER 6
Intelligence and the Problems of Society

School Dropout rate, under-employability and other societal problems can be greatly reduced

The tapestry of this book needs to include a representation of how intelligence relates to the problems of society such as school dropout rate, under-employability, criminality, and welfare. The brighter and the more pronounced the threads describing intelligence, the less vivid each of the threads representing problems of society appears in the tapestry; problems of society decreasingly dominate the tapestry as the fabric is increasingly overwhelmed by the more brilliant threads of intelligence.

A middle-school student that I tutored and mentored, though quite bright, was also a very angry child. He was thought to be a problem child by the administration and was viewed as someone who potentially might become much worse. He was also the school tough. I remember walking one day with him down the school's main hallway while children were moving from one class to another. For those who do not recall, moving between classes in middle-school during the five or so minutes allowed to reach one's next classroom, and to do that with near-teenage children, is a challenging experience. However, unlike other children, who weave quickly among each other from one classroom to another in apparent chaos, for this child the seas parted; everyone kept out of his way. He walked down the center of the hallway with no one touching him.

When arriving at our destination, and being alone, I made comment of what had just happened. I told him being the school tough might seem powerful, but it does not make friends. It sets you apart. Even your apparent friends are not likely to be real friends. We talked at length. And when I say we talked, I mean we talked about it openly and honestly. I did not tell him what to do as a parent or teacher might, appropriately. We discussed the problem with direct, sincere, and uncomplicated feeling. It made a difference. A tutor-mentor sometimes can develop a powerful, meaningful, and in many ways a differently productive relationship; incidentally, one that is all-too-often inadequately appreciated by teachers, but frequently makes an enormous and long-lasting difference. I talked of this child in an earlier section in relation to advocacy and showed how this child had significant potential and subsequently found his way to success.

We have all experienced good teachers and not so good teachers. Our son, while still in kindergarten, was asked by a second-grade teacher to play something on his violin for her class. He declined. She then asked him whether he would just show the instrument to the class. He allowed that that was okay. She then asked him to show the class the bow, which he did. Then she asked him to play a note, just one note, and he did. And finally, she asked him to play just one piece which he now felt comfortable to do. That was an exceptionally good teacher.

On the other hand, in high school our daughter was honored with an invitation to attend the Governors School for the summer to study the clarinet as a member of the orchestra. She quickly accepted but then was told by the principle of her high school that she could not attend because she had to take gym during summer even though she was on the tennis team during the spring season. That was exceptionally bad administration.

There are many good teachers and, of course, some that are not so good. All of them have a powerful influence over our children that surely is less than adequately appreciated.

We know that intelligence can be enhanced, most substantially among the disadvantaged. Yet, over the years, many respected researchers have taken the opposite position; that intelligence is essentially stable and can hardly be changed. This position, while somewhat valid for IQ scores, is not valid for intelligence; while somewhat true for the advantaged is much less true for the disadvantaged. The following comments suggest real concern with this apparent dichotomy of views. Jensen, for example, conveyed the view that intelligence is stable in his 1969 article in the *Harvard Journal* entitled *"How much can we boost IQ and scholastic achievement?"* This stirred extensive debate and discussion, but little resolution.

Twenty-five years later, in their 1994 book, *The Bell Curve*, Herrnstein and Murray incredibly expanded on this thinking, writing,

> The cognitive elite, with its commanding social position, will implement an expanded welfare state for the under-class [those with relatively low intelligence] that also keeps them from underfoot. Our [, Herrnstein and Murray's,] label for this outcome is the custodial state[103] ... a high tech and more lavish version of the Indian reservation for some substantial minority of the nation's citizens, while the rest of America tries to go about its business.[104]

They predict the need of a two-tier society. This kind of prediction is enough to make one's hair stand up – certainly, enough to make one sit up and take notice.

Notably, Herrnstein and Murray indicate that the problems are not so much caused by people with low socio-economic-status but by people with low intelligence. Setting aside, for the moment, the validity of this observation, and assuming it is correct, Herrnstein and Murray's focus is to separate those people from society as hopeless, whereas here in this book, the focus is to raise their intelligence and integrate them into society; importantly, we know how to do that.

But they also make another claim that is simply groundless, namely that intelligence can hardly be modified. Herrnstein and Murray write, "the story of attempts to raise intelligence is one of high hopes, flamboyant claims, and disappointing results. For the foreseeable future," they continue, "the problems of low cognitive ability are not going to be solved by outside interventions to make children smarter" (p. 389). The overwhelming evidence documented in the literature and reviewed in this book belies these comments. Incidentally, in 2012 Murray writes in another book, *Coming*

---

[103] *The Bell Curve*, p. 523
[104] *The Bell Curve*, p. 526

Apart: The State of White America, that the problems, including economic inequalities, will only get worse.

The Herrnstein and Murray predictions are based on two underlying claims, the first being accurate; the data shows that problems of society are currently disproportionately associated with people having low IQs. The second claim, suggesting that intelligence cannot be meaningfully enhanced, is simply unfounded. We first focus on Herrnstein and Murray's accurate message. While their data indicates that the problems of society are disproportionately associated with those with the lowest IQs, and though no relationship is indicated, a causal relationship is rather strongly implied when they write, "For most of the worst social problems, the people who have the problem are heavily concentrated in the lower portion of the cognitive ability distribution" (p. 369). With unemotional analysis this associative evidence is rationally shown to be causal, as seen shortly.

This awareness has far reaching ramifications on each individual and society. This chapter submits, with extensive detail, that the problems of society may be substantially reduced as intelligence is increased.

So, in this chapter we review the data concerning the relationship of the number of people having IQs in the lowest 20 percent with various problems of society including school dropout rate, employability, criminality, poverty, chronic welfare, frequency of LBW babies, illegitimate children, and automobile fatalities. A source of this data is the book, *The Bell Curve,* yet, curiously, the conclusions reached in this book are in direct conflict with that book's central message. Incidentally, while the data is several years old, the data is essentially unchanged at this writing.

Importantly, the results showing a substantial reduction in the problems of society only touch the tip of the iceberg – they indicate only the surface message. The more profound effects are the changes on each individual involved, their self-satisfaction, their pride in achievement, and their personal perceived contribution to *their own* health and wealth. And this, of course, extends to the *entire* society's health and wealth because, as we have learned throughout the world, as more people participate in the economic life of the society, the economic advantages for all increases; as mentioned earlier in this book, with more people participating in the fruits of society, demand for goods and services increases, revenues increase, those that provide the goods and services benefit – all members of society, the disadvantaged as well as the advantaged, benefit substantially.

In stark contrast with the major message conveyed by Herrnstein and Murray, intelligence changes with every experience and can be significantly enhanced with environmental improvements, including in health, nutrition, and general well-being. This is desirable not only to permit people to more closely reach their potential but also to reduce precisely those societal problems such as school dropout rate, under-employability, and criminality, asserted by some as being unavoidable. Evidence presented throughout the literature and collected and expanded in this book suggests that intelligence is significantly modifiable starting during prenatal care and according to each of the subsequent environments of the many developmental stages of life. The expense for these improvements, such as providing vitamin and mineral supplements, is often trivial, often show immediate returns, and are essentially immediately and entirely offset by a decrease in related societal costs. Whatever the price, nutritional enhancements and other environmental improvements are a lot less costly than the societal consequences of viewing intelligence as being fixed and predetermined.

It is predicted here that we can maintain a society with not only equal rights under the law, but also, improved opportunities for all, more effective educational opportunities, and improved job opportunities—improved opportunity to pursue the benefits of society. Far from incorrectly highlighting society's disturbing eventuality of a two-tiered society, this book, utilizing the same data presented by Herrnstein and Murray, forecasts a single-tier integrated society with not only dramatically reduced problems but increased opportunity for all. By focusing on the means to effectively and efficiently improve intelligence, especially of the disadvantaged, the problems of society are substantially reduced.

The costs of the problems of society take several forms. The monetary costs alone are enormous, amounting to more than a trillion dollars per year—staggering. For example, the annual cost to maintain the current 2.3 million prisoners in American jails is $92 billion. This cost, of course, does not include the costs to society these prisoners had perpetrated on society, or the immense security expenses such as police to limit continuing crime. Annual unemployment payments to the millions of unemployed amounts to about $100 billion.[105] This cost does not include the loss of tax revenue if these people were working in the community. And, of course, the more unemployed the more costs for food, social services, as well as medical costs. Additionally, state and federal welfare payments amount to about $600 billion per year. The cost of welfare, alone, considerably exceeds the budget for national defense (see graph below, Figure 13). (These data do not reflect the Iraq or Afghanistan wars.) And these are just a few of the many expenses to society due to the problems of society. Furthermore, the cost to each individual in society, those directly involved, and those indirectly affected, is hardly measurable.

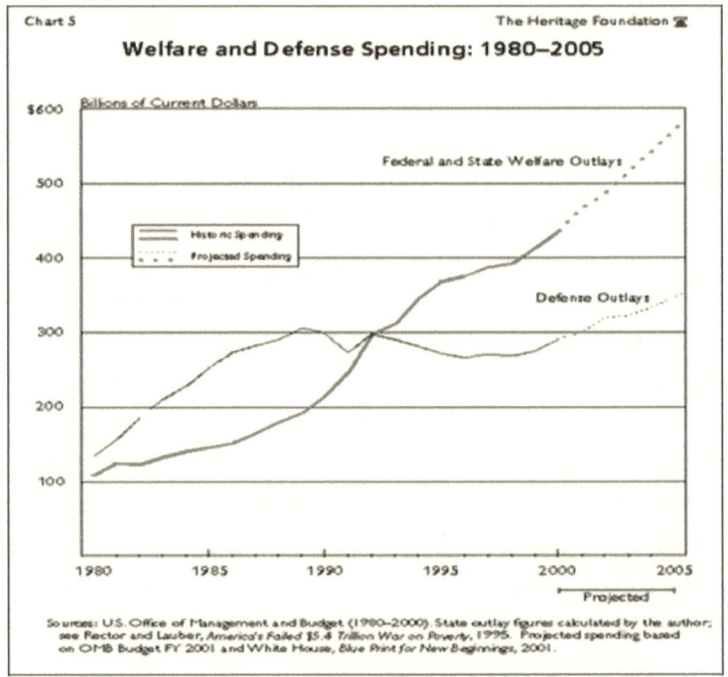

Figure 13

Specifically, the data shows that approximately 60 percent of the problems of society such as poverty, prison population, school dropout rate, and under-employability, are associated with people whose IQ is among the 20

[105] State and federal unemployment insurance programs have cost roughly $520 billion, for the five years from 2008 to 2012, according to the Congressional Budget Office.

percent lowest IQ scores (see following chart[106], Figure 14). The extensive evidence, though associative, is clear; the more children that dropout of school, the greater the number of under-employable, the more welfare. (While associative, this evidence will be shown to be causal, as well, within this chapter.) The chart shows that as intelligence increases the problems of society decrease. This assertion is one of the most important messages of this book.

This places a much more immediate, a much more dramatic, impetus on understanding intelligence, and understanding the relationship between intelligence and poverty as well as between intelligence and school dropout rate, employability, and various other problems of society; it significantly addresses many ways to gradually yet substantially reduce not only poverty but also other problems of society. Of course, while the resulting potential economic savings are substantial there are many other cost savings resulting from reduction in poverty and other problems of society felt not only by those directly affected but also those indirectly effected.

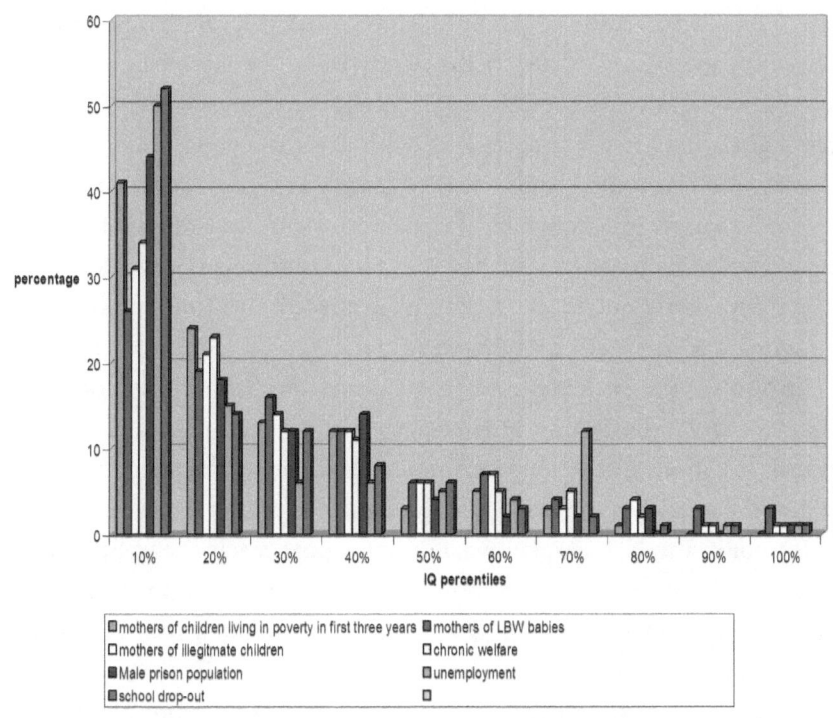

Figure 14

---

[106] This chart was also shown in Chapter 2. References are listed with that chart.

In particular, 66 percent of school dropouts have an IQ in the lowest 20 percent (quintile). As intelligence is raised for these young people, via enhanced nutrition, improved health, and thus more effective learning experiences, their interest in school activities will likely increase, their success in school will probably increase, and they will more likely complete school. Incidentally, 74 percent of prison inmates are school dropouts. Raising intelligence of those toward the lower end of the intelligence spectrum will likely disproportionately reduce prison populations. Similar observations may be made for under-employability, poverty, and the occurrence of LBW babies. The central message of this chapter is that the staggering human pain, injustices, and societal costs just mentioned can be significantly reduced, amazing as it may seem, by raising the intelligence, as measured by IQ tests, of those toward the lower end of the intelligence spectrum, and this can be done in known ways, as has been shown earlier.

Furthermore, providing such environmental modifications is not costly and may be entirely offset by related savings. Vitamin and mineral supplements, provided to the disadvantaged, could result in large change for trivial, if any, expense. Much of the cost is nearly immediately offset by reduced costs elsewhere, also discussed earlier. Moreover, the costs of such environmental enrichments when compared with the cost reduction of problems in society, such as under-employability, prison expenses, and welfare costs, prove to be a bargain, not to mention the raised self-satisfaction of the many people involved. Of course, mineral and vitamin supplements are only part of the story; better nutrition, in general, is needed within a supportive, encouraging, and motivating environment.

To be clear, the evidence used here is associative data; it does not directly present a causal relationship. However, upon applying a minimal amount of rationale a causal relationship surfaces; abundant smoke suggests there is fire. Let me quickly explain. While the data regarding school dropout rate and IQ, for example, is associative, it is not a great reach to accept that individuals who have low IQs are likely to have difficulty attending to, absorbing, and retaining formal and informal learning experiences; they are likely to fall successively further behind in class-work; they are likely to become increasingly unable to follow the material covered and to become bored – to be easily distracted; they are likely to increasingly dislike attending school and when possible, will tend to drop out of school. There appears to be a causal relationship between school dropout rate and low IQ, yet the specific data presented is associative. And the opposite rationale is also true; as intelligence is raised, individuals will be more able to attend to formal and informal learning experiences, will do better in school, keep up, feel success in school, and will more likely complete school. The questions of causal relationship of the data are further discussed toward the end of this chapter. Furthermore, there is a strong relationship between the various problems of society such as employability and school dropout rate; a person who dropped out of school will more likely be less employable, and more likely to be on welfare. Suffice it to say, here, that rational analysis, just presented, indicates that while the relationship is technically associative, there is reason to claim it to be causative.

As societal advantages are more widely enjoyed, and as ample nutrition, good health, and effective formal and informal learning are more broadly experienced, intelligence is likely increased for many, especially those currently near the low end of the intelligence spectrum. Many will, thus, more readily experience more of the health and wealth of society. They will tend to protect their health and wealth, and thus, the problems of society would be reduced. It will improve the potential tranquility and stability of society and thus enhance its economic and social advantages.

While the extensive evidence showing the strong association between the problems of society and intelligence is not questioned, the assumptions, as presented by Herrnstein and Murray, that limited intelligence can hardly be changed and their pessimistic predictions that the problems of society cannot be reduced, are entirely rejected.

The book *The Bell Curve* conveys these statistics in detail, and concludes, as did Jensen twenty-five years earlier, as do many psychology professionals, that intelligence is largely determined by genetic factors and is relatively fixed. Try as one might, they claim, there is not much potential to enhance intelligence. These observations are far from isolated. In 1987 Snyderman and Rothman surveyed 1,020 professionals of various fields such as psychology, genetics, cognitive sciences, education, and psychometrics. These people were asked whether genetics or the environment plays a stronger role in determining intelligence. About 21 percent felt unqualified to answer, while the remaining 79 percent were essentially evenly split as to whether there was enough information to make a judgment. Of the 79 percent who felt there was sufficient information, about half, 39 percent concluded "that 60% of the variation in IQ within American white population is associated with genetic variation". Though this study is about 30 years old, some academics, such as Rushton, have cited this survey as evidence of the validity of this position. Rushton cites this Snyderman and Rothman survey highlighting that the racial gap in IQ is likewise believed to be caused significantly by genetic factors.

Thus, people with low intelligence—*according to these professionals*—hardly modifiable low intelligence—will continue to perpetrate the problems of society. The problem, many claim, can hardly be corrected. In fact, some add, the problems will only get worse because those with low IQ scores tend to have more children, and so society will increasingly be disturbed by the conflicts between the IQ haves and the IQ have-nots. Also, they claim that the haves will tend to marry haves and the have-nots will tend to marry have-nots exacerbating the problem. This is a disquieting and troublesome prediction. Fortunately, while possibly viewed by some as increasingly true, it is not so.

These pessimistic notions simply do not hold upon close analyses. They demand re-examination and rational and unemotional scrutiny. This book reveals the weaknesses of previous analyses and, using the identical data as used in those previous studies, supports just the opposite conclusions.

What would explain these statistics? Why do those with the lowest IQ scores disproportionately have poor health, disproportionately drop out of school, are more persistently out of work, tend toward crime, or disproportionately have LBW babies? The answers to these questions may seem too apparent and too easy; *raise the IQ scores of at least those with the lowest IQ scores and the problems of society would be significantly reduced.* Can this apparently simplistic answer address such historically complex sociological issues? Surely, there are other related factors to be discussed as we proceed, but not so incredibly, as will be shown immediately, collectively this seemingly simplistic response *does* answer a significant part of the problems.

Before proceeding, several additional observations need be made, and a few need to be reemphasized. While some studies report the effects of individual environmental factors on intelligence, and others report the influences of multiple influences, little specific information conveys the extent of influence of any one factor in combination with any other one factor. It is thus difficult to unequivocally claim how additive some factors are regarding others. However, rational analyses permit one to conclude that while informal and formal learning, health, and nutrition are not independent, their influences on intelligence are significant. Furthermore, while incremental experiences in each of these areas of learning while having adequate nutrition are largely additive, they likely reach a level of diminishing returns. For example, giving mineral and vitamin supplements to a well-nourished individual hardly results in increased intelligence.

Also, studies do not demonstrate the extremes of environmental influence. For example, a starving individual behaves less intelligently than that same person when merely poorly nourished, not to mention when well nourished. For ethical reasons, studies do not reflect the extremes of environmental conditions. No conclusions can be made with the current information as to the full scope of environmental influence on intelligence. The information, though, does overwhelming indicate that the environment greatly influences intelligence.

The issue of whether differences in intelligence between groups of people are even partly genetically determined will be revisited in the next two chapters, but regardless of one's view of this issue, there can be no doubt that especially among the disadvantaged, human intelligence is extensively determined and often severely inhibited by the limitations in the environment experienced. Additionally, though numerous studies have provided abundant information about the effect of specific environmental factors, few have ventured an estimate of long term and collective environmental influence on intelligence. Several studies mentioned earlier reflect various long-term environmental influences (for example, Scarr and Weinberg, 1976; the Smilansky study, the Milwaukee Project, 1971, and Bronfenbrenner, 1974). These studies suggest that environmental influence can lead to enhancements of IQ by between fifteen and thirty points. I hasten to add, this is not to suggest that the average individual's IQ may be raised by thirty points. Furthermore, these as well as the vast majority of studies, focus on young children, not adults.

So, while both genetic factors and environmental factors determine intelligence, environmental influences support or limit the expression of genetically influenced capabilities. Essentially every human experiences periods of illness, limited nutrition, or emotional stress. During these time periods, environmental influences effect the neuronal system's well-being. *Among the disadvantaged, these time periods are likely much more frequent and extensive; expression of genetic factors regarding intelligence, and the well-being of the neuronal system are, likely, more inhibited because of more limited nutrition and health, and accordingly, also less effective accumulation of information from both formal and informal learning experiences. And the opposite is true for the advantaged; the expression of genetic factors regarding intelligence and the well-being of the neuronal system are less inhibited because of likely adequate nutrition and health, and accordingly, more effective accumulation of information from both formal and informal learning experiences.*

We can now relate several conclusions namely, (a) intelligence of each of us may be significantly enhanced, (b) intelligence test scores of an individual at age six or sixteen convey limited information about that individual's potential, and while it may be reasonable to predict that a person who appears intelligent, or unintelligent, at a given moment, is *likely* to remain so, it is *not* reasonable to predict that a person who appears intelligent, or unintelligent, at a given moment in time is *destined* to remain so, and (c) intelligence test scores, whether high or low, indicate current behavioral characteristics, and accordingly, indicate previous and current environmental conditions; they unreliably predict future intelligence; while IQs appear stable with stable environments, IQs are not stable with changes in environments as suggested here.

For some, these implications are difficult to accept. They challenge the perception of our own experiences. They contradict what we have been led to believe, over the years, by experts; they challenge our belief system. They conflict with school system directions and they dispute society's treatment of groups of individuals. They cause us to question our feelings about others and most importantly about our children and ourselves. These observations make us uncomfortable, causing us to doubt and question, and that is precisely the intent.

The two-tiered society, predicted by the authors of *The Bell Curve*, consisting of IQ haves and IQ have-nots with its anticipated conflicts, need never occur. Surely, there will always be people with relatively low intelligence. And while the percentage cannot be reduced, their intelligence, at least as measured by IQ tests, can be significantly improved and thus their effectiveness and productivity in society can be dramatically enhanced.

Several of the problems of society are now discussed in turn. As will be clear from the data, people with low IQs are disproportionately associated with these problems of society. First shown, for each area of concern, are the data highlighting the concern.

The overall high school dropout rate in the United States in the year 2012 was about 20 percent; for whites the dropout rate was 14 percent; for blacks it was more than twice that at 31 percent, and for Latinos it was 27 percent. Taking an additionally useful cut, for the economically disadvantaged the dropout rate was 28 percent[107]; more than one out of four disadvantaged children never completed high school. While appearing to be a racial concern, predominantly, the data identifies the problem as related to being disadvantaged. These data reveal a profound problem for the country, but perhaps, more importantly, for each of the individuals and their effected families. These dropouts are likely to be severely inhibited from participating in society's benefits. These people not only lose out in not being educated, but they are more likely to find themselves in prison; 74 percent of state prison inmates are school dropouts.[108] Also, school dropouts are more likely to be unemployed and be less employable and make much less money than others when they do work.

As shown in previous sections of this book, investments for environmental improvements for those at the low end of the economic spectrum would be overwhelmingly rewarding, not only to the national treasury but more importantly to each individual involved. Having said that, I must remind the reader that the costs of these environmental enhancements will be more than offset by the *immediate* reduction of other related costs in society. These environmental improvements would likely raise intelligence, interest in learning, success in school, and motivation to remain in school through graduation, thus reduce the high school dropout rates.

Expanding on this, the following chart, figure 15, shows that those who complete high school earn annually about $9,000 more than those who do

---

[107] Public High School Four-Year On-Time Graduation Rates and Event Dropout Rates: School Years 2010–11 and 2011–12. First Look **April 2014 Marie C. Stetser Robert Stillwell** National Center for Education Statistics
U.S. Department of Education **NCES 2014-391**

[108] More specifically, about 74 percent of America's state prison inmates, almost 59 percent of federal inmates, and 69 percent of local jail inmates did not complete high school (Harlow, 2003); Harlow, C. W. (January 2003). Education and correctional populations (NCJ 195670). Retrieved from Bureau of Justice Statistics, US Department of Justice web site http://www.bjs.gov/content/pub/pdf/ecp.pdf

not complete high school. Those who complete college make about $20,000 more per year than those who just complete high school. The specific numbers will change but the relative disparities may likely increase. Importantly, not everyone should, or wants to, attend college; some, more appropriately prefer to develop a skill or talent.

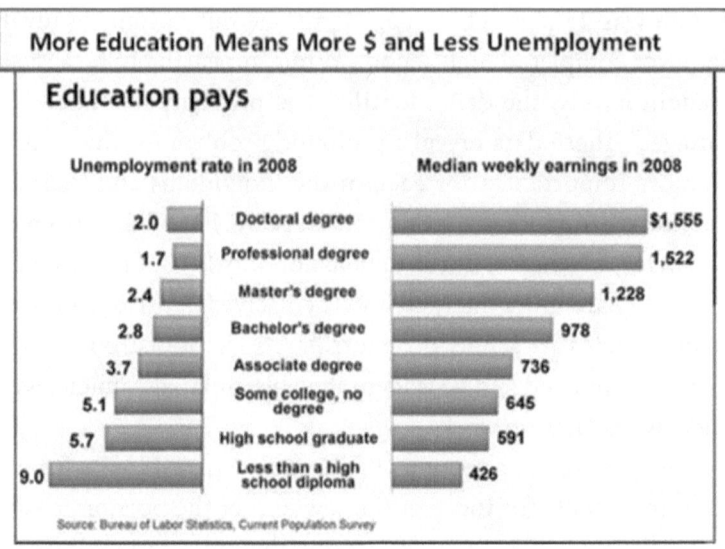

Figure 15

On a related subject those who complete college pay significant taxes every year and across their lifetime whereas many of those who do not complete college pay little if any income taxes across their lifetime. The individual who stays in school through high school is significantly rewarded for his or her efforts. Those who stay in college through graduation often recognize a relatively large return for the investment. Schooling provides financial rewards for the individuals involved as well as the government that helped support the activity but perhaps much more to the point, for many, formal learning experiences may provide more satisfaction in one's life activities than the alternatives.

In some areas of the country, the school dropout data is more dramatic than others. The percent of students that do not graduate from high school in the largest cities is much higher than the overall country average. Notably, "Eighty percent of high-school dropouts are boys and less than 45 percent of students enrolled in college are young men."[109] In addition, while forty years ago, black boys were about twice as likely as white boys to drop out of school (a ratio 10 to 5) over the years the ratio has dropped to about 10 to 6[110]. This is partly explained by the drop in the black/white IQ gap in the same period, from about fifteen points to about ten points.[111] The following chart, figure 16, expands on these comments.

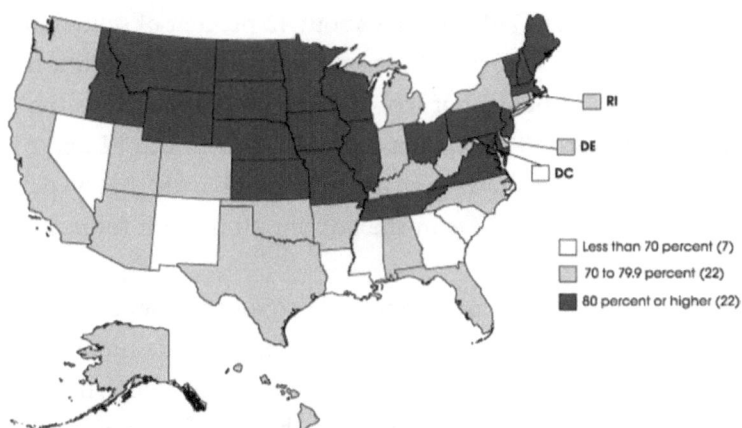

Figure 16. Averaged Freshman Graduation Rate (AFGR) for public high school students, by state or jurisdiction: School year 2009–10

NOTE: The Averaged Freshman Graduation Rate is the number of graduates divided by the estimated freshman enrollment count 4 years earlier. This enrollment count is the sum of the number of 8th-graders 5 years earlier, the number of 9th-graders 4 years earlier, and the number of 10th-graders 3 years earlier, divided by 3. Ungraded students are allocated to individual grades proportional to each state's enrollment in those grades. Graduates include only those who earned regular diplomas or diplomas for advanced academic achievement (e.g., honors diploma) as defined by the state or jurisdiction. Race categories exclude persons of Hispanic ethnicity. Total includes students for whom race/ethnicity was not reported or whose race/ethnicity is not represented in the five racial/ethnic categories presented in this figure.

SOURCE: U.S. Department of Education, National Center for Education Statistics, Common Core of Data (CCD), "State Dropout and Completion Data File," 2009–10. See Digest of Education Statistics 2012, table 124.

Figure 16. Sources and other information

---

[109] *Newsweek,* September 19, 2005, p. 59

[110] US Department of Education, Institute of Education Sciences, National Center for Education Statistics

[111] W. T. Dickens and J. R. Flynn, (2006), *Psychological Science 17(10),* p. 924.

So, the problem is regional, racial, and cultural. It has to do with socio-economic status, but as pointed out by many professionals (for example, Herrnstein and Murray in *The Bell Curve*, 1994) and as will become apparent immediately, it is largely about IQ; those toward the low end of the IQ spectrum disproportionately drop out of school. They are far more likely to drop out of school than those who are in the high end of the spectrum. See the following table, figure 17, using data shown in *The Bell Curve* (p. 372). The line chart of figure 17 shows that about 42 percent of school dropouts have an IQ in the lowest decile, the lowest ten percent (IQ below 81), and, additionally, about 24 percent of school dropouts have an IQ in the second lowest decile, the second lowest ten percent (IQ between 81 and 87). The accompanying pie chart shows perhaps more graphically that 66 percent of high school dropouts have IQs in the bottom 20 percent (quintile) of the IQ spectrum.

But while this data suggests the school dropout problem is significantly due to intelligence, one might observe that this appears too simplistic; surely, many different factors influence children to drop out of school, including unexpected pregnancy, family economics, death in the family, and emotional situations. But often these are not the fundamental causes. Unquestionably, all these other factors surely play a role, however, the data points most predominantly to intelligence. Lack of interest in school, education perceived as lacking relevance by the child or parents, but, most importantly, low intelligence, likely, explains much, if not the clear majority, of school dropouts.

Again, while appearing simplistic, the rationale is clear. IQ scores predict school success. Incidentally, this was the motivation for the creation of Binet's first IQ test. Raising IQ scores improves the likelihood to experience interest, capability, and thus success in school. This, in turn, raises the likelihood to enjoy school, and thus the likelihood to stay in school to completion.

Two-thirds of high school dropouts had IQs in the bottom two deciles of the spectrum

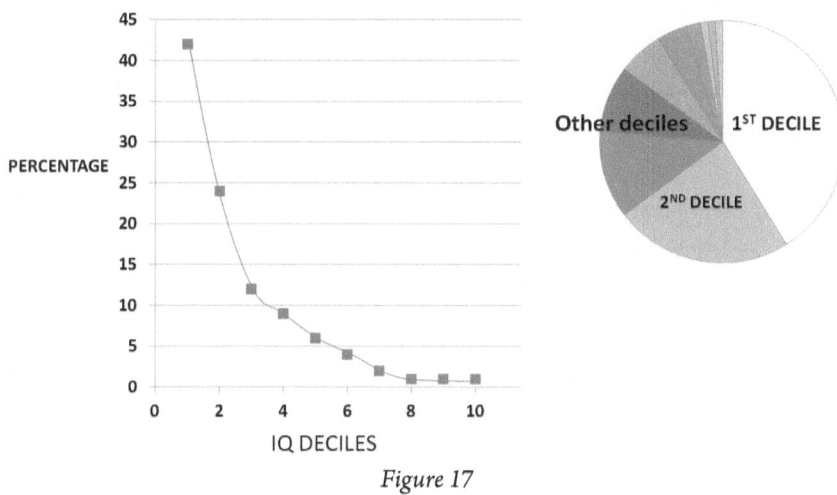

Figure 17

By way of a specific example, I was asked to tutor and mentor a sophomore in high school who was about to dropout. He hated school. Every day, upon coming home after school, he would throw his books into a corner and take a shower for half an hour to wash the filth of school off his body. At our initial meeting, the boy's mother asked me why I would make a difference when all the previous special attention did not work. I had no idea how to answer this straight forward question. I said, we'll be working one-on-one and that might make the difference. I worked with the boy, we'll call Bill, for some time and gradually he became comfortable in school, comfortable enough to where he felt that he could graduate with a certificate of completion, and was told so. With a little more time and effort, it became clear that this was not enough for him. After about two and a half years of consistent caring and increasing successes, Bill graduated, on time, with a full diploma. His pride and self-satisfaction overwhelmed all those around him—his teachers, parents, sibling, and, of course, himself.

This child had been labeled and treated as learning-disabled through elementary and middle schools. His need was not nutrition or improved well-being. The problem was not cognitive disability. It was lack of positive experiences essentially throughout his school career which caused a lack of self-confidence, and a feeling that school was not worthwhile, interesting, or fun. More particularly, being placed in classes for the learning-disabled is all-too-often viewed by children as being punished and now separated from their peers; that they are not smart enough to be in regular classes. While really learning-disabled children need special attention, many others, are misidentified. All these children in both groups must be differently and appropriately treated with sensitivity.

We know how to significantly raise intelligence, at least as measured by IQ tests, by an average of ten points which would raise the IQs of most of those between 55 and 86 to above the 20-percentile level of 87.

Most school dropouts, 66 percent, have IQs in the lowest 20 percent. Imagine raising intelligence and the corresponding benefits, such that motivation, expectations, and self-confidence are raised as well. With improved nutrition and other environmental factors comes greater intelligence, at least as measured by IQ tests. These students would, more likely, achieve more success in school, more likely enjoy school, and thus more likely complete school. The IQs of most of these students, whatever their race or socio-economic status, could be raised above the 20th percentile by raising their IQs by an average of about ten points with improved nutrition, health, and effective learning experiences. The number of dropouts would be significantly reduced. Many would likely join those that complete school and thus have better job opportunities and join the members of society who more fully enjoy its wealth and health.

Unemployment

Societies experience different levels of unemployment according to a myriad of factors. Of course, one of these is the basic economic conditions of the society. Additionally, of course, employment of each individual depends on capability, education, health, location, and other factors. The following chart, figure 18, shows unemployment rates according to level of education.

FIGURE 4

U.S. Unemployment Rates across Education Groups (Ages 25+)
2000:Q1 - 2010:Q4

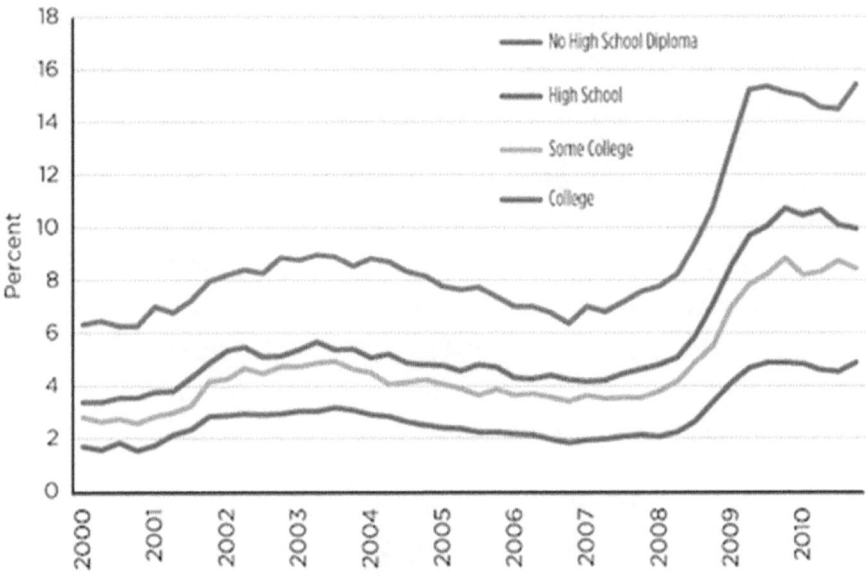

SOURCE: Bureau of Labor Statistics/Haver Analytics.

Figure 18

The important observation, here, is that the ratios of unemployment regarding educational level remain about the same in good and bad times. The various curves of the chart show the relationship between unemployment and schooling. The top curve shows the unemployment rate for high school dropouts, the curve second from the top shows unemployment for those who completed high school, the next curve shows unemployment rates for those who had some college, and the lowest curve shows unemployment rates for college graduates. Notably, some people in the society are consistently far more prone to being unemployed than others regardless of the economic conditions.

Thus, unemployment is partly due to economic conditions but also substantially due to educational limitations and capability limitations. These are directly related to low IQ. As shown in figure 19, the lower the IQ the more likely to drop out of school, the more likely to be unemployed.

Intelligence, or more specifically, here, IQ score, is one of the key factors that explain unemployment. The following line chart of figure 19, conveying data shown in *The Bell Curve* (p. 374), shows that those with low intelligence are disproportionately unemployed. The line chart of figure 19 shows that about 50 percent of men who did not work in 1989 had an IQ in the lowest decile, the lowest ten percent (IQ below 81), and about 14 percent of men who did not work in 1989 had an IQ in the second lowest decile, (IQ between 81 and 87). The accompanying pie chart shows more graphically that 64 percent of men who did not work in 1989 had an IQ in the bottom 20 percent of the IQ spectrum.

Blacks are far more likely to be unemployed than are whites. That is partly because they have, on average, a lower IQ, which, as shown earlier, is essentially due to being disproportionately disadvantaged. The low IQ also partly explains why blacks are far more likely to have dropped out of school, as shown above.

Moreover, people with less than an IQ of 80 cannot, by law, be considered for the armed services. In practice, the armed services do not consider those with an IQ below 85. That has prevented nearly half the black population from being considered for the armed services. With a reduction of environmental constraints, this problem would be greatly reduced, if not essentially eliminated.

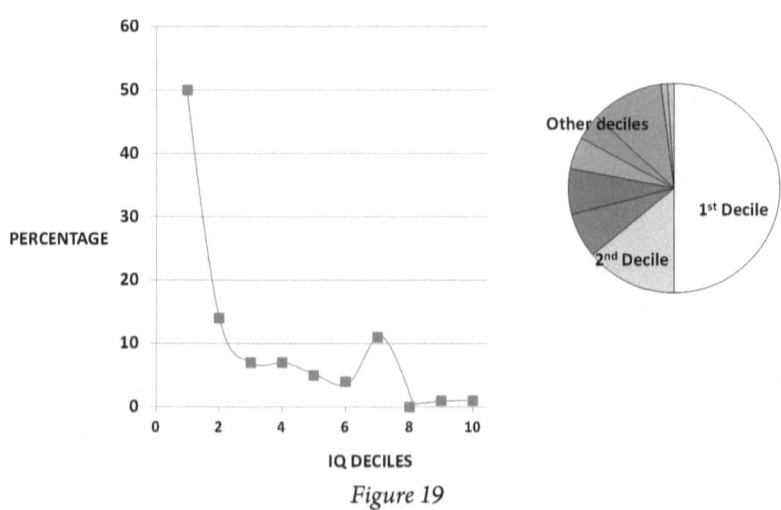

Sixty-four percent of able-bodied men who did not work in 1989 had IQs in the bottom 20 percent of the IQ spectrum

Figure 19

These are not racial explanations for disproportionate black unemployment, though there are some. For example, studies have shown that when two candidates with essentially equal qualifications apply for a job, one of them black, the non-black is likely to get more consideration. To be sure, there are clear racial prejudices that effect employment, prison population, and other problems of society.

The objective of this book is to convey that by enhancing intelligence, especially of the disadvantaged—whether black or white—capability will increase, opportunities will increase, and economic inequalities will decrease. And while this book does not focus on the racial prejudice issues directly, or racial injustices directly, we do address, in this author's view, the single largest obstacle to black equal employment opportunities, that being environmental constraints on developing one's intelligence.

Reviewing the chart above, it appears that by raising IQ scores, of those in the lowest quintile of the IQ spectrum, by an average of only ten points[112] would significantly reduce unemployment and other related problems of society. It would move most of those currently below the twenty percent mark to above that mark, and to the much lower rates of unemployment. And while it would benefit all people, it would especially benefit the disadvantaged, the blacks, and most especially young black males.

Crime rate and prison population

The incarcerated population in the United States has grown over the years for a variety of reasons. The following chart, figure 20 shows this growth between 1920 and 2012 increasing most sharply starting in the early 1980s.

---

[112] Raising IQs by an average of ten points of those having IQs below 87, the 20 percent mark, would raise the IQs of essentially all to be above 87. This is so because those towards the low end of this spectrum are few in number, whereas those towards the high end of this spectrum are great in number, and, because the lower the IQ the more likely, and the more readily, it is to be raised. This was described more fully earlier in the book.

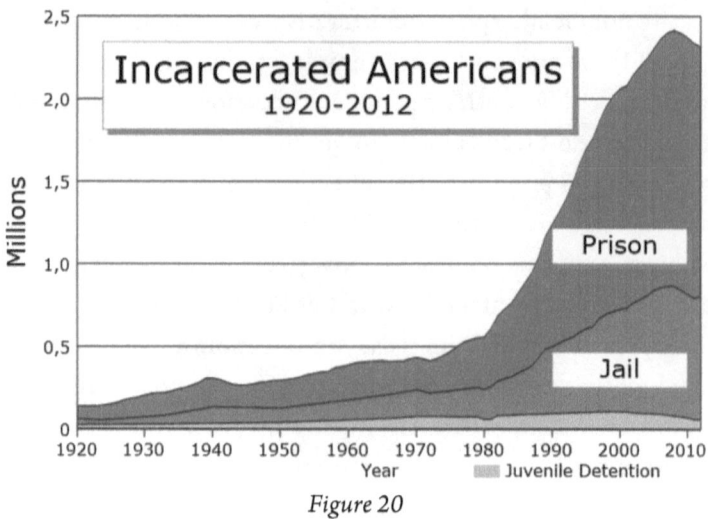

Figure 20

According to the U.S. Bureau of Justice Statistics (BJS), 2,266,800 adults were incarcerated in U.S. federal and state prisons, and county jails at year-end 2011 – about 0.94% of adults in the U.S. resident population. Additionally, 4,814,200 adults at year-end 2011 were on probation or on parole. In total, 6,977,700 adults were under correctional supervision (probation, parole, jail, or prison) in 2011 – about 2.9% of adults in the U.S. resident population. A 2014 report published by the National Research Council asserts that the prison population of the United States "is by far the largest in the world. Just under one-quarter of the world's prisoners are held in American prisons."[113]

Several important observations are to be made from this and the following chart. First, the incarceration rate in the United States has increased in the last generation nearly four-fold to where it is now the highest rate in the world. Remarkable! (The US Justice Department has addressed this, but the current administration is reversing much of the previous administration's modifications.) Second, in 2006 the percentage of prisoners by race and ethnicity is shown in the following chart; in percentages, nearly 2½ times as

---

[113] http://en.wikipedia.org/wiki/United_States_incarceration_rate

many Hispanic as whites are incarcerated, and seven times as many blacks as whites are in prison.

According to Charles Street,[114]

> United States Bureau of Justice Statistics director Jan Chaiken reported 30 percent of African-American males ages 20 to 29 are "under correctional supervision" either in jail or prison or on probation or parole. Especially chilling is a statistical model used by the Bureau of Justice Statistics to determine the lifetime chances of incarceration for individuals in different racial and ethnic groups. Based on current rates, it predicts that a young Black man age 16 in 1996 faces a 29 percent chance of spending time in prison during his life. The corresponding statistic for white men in the same age group is 4 percent. ... Nothing can excuse policymakers and activists from the responsibility to end racist criminal justice practices that are significantly exacerbating the difficulties faced by the nation's most truly and intractably disadvantaged. More than merely a symptom of the tangled mess of problems that create, sustain, and deepen America's savage patterns of class and race inequality, mass incarceration has become a central part of the mess.

Why the huge incarceration rate and why are black males and Hispanics disproportionately incarcerated, as seen in the following chart, figure 21?

---

[114] Originally published in *Z Magazine*

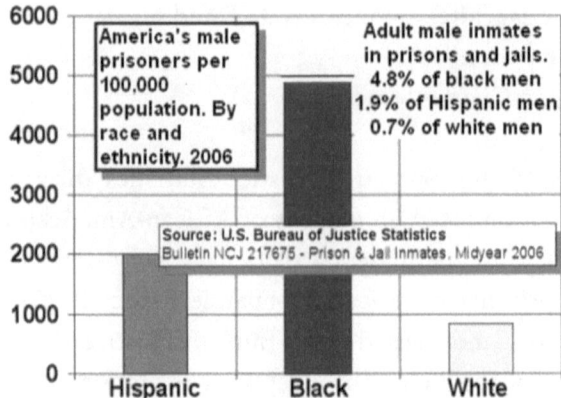

Percentages of black and white men exclude Hispanic men.

Prison and Jail Inmates at Midyear 2006 (NCJ 217675). U.S. Bureau of Justice Statistics. Male incarceration by race. The percentages are for adult males and are from page 1 of the PDF file: "On June 30, 2006, an estimated 4.8% of black men were in prison or jail, compared to 1.9% of Hispanic men and 0.7% of white men."[115]

Figure 21

Though the answers and resolutions are less than clear, the sociological implications are great.

And even those data do not tell the full story. The following two charts, figure 22 and 23 add more relevant details.

---

[115] http://en.wikipedia.org/wiki/United_States_incarceration_rate

Year	White	Black	Other
1986	2,090,100	1,117,200	32,100
1987	2,12,200	1,231,100	36,300
1988	2,348,600	1,325,700	39,800
1989	2,521,200	1,489,000	45,400
1990	2,665,500	1,632,700	49,800
1991	2,742,400	1,743,300	49,900
1992	2,835,900	1,873,200	53,500
1993	2,872,200	2,011,600	60,200
1994	3,058,000	2,018,000	65,300
1995	3,220,900	2,024,000	90,200
1996	3,294,800	2,083,600	104,500
1997	3,429,000	2,149,900	113,600

Number of adults in prison or jail, or on probation or parole by year and race

Source: Bureau of Justice Statistics Correctional Surveys (The National Probation Data Survey, National Prisoner Statistics, Survey of Jails, and The National Parole Data Survey) as presented in Correctional Populations in the United States, 1997.

Figure 22

The following chart relates the data in percentages which may be more revealing.

Year	White	Black	Other
1986	1.4%	5.7%	0.6%
1987	1.4%	6.2%	0.6%
1988	1.5%	6.6%	0.7%
1989	1.6%	7.3%	0.7%
1990	1.7%	7.6%	0.7%
1991	1.7%	8.0%	0.7%
1992	1.8%	8.5%	0.7%
1993	1.8%	9.0%	0.8%
1994	1.9%	8.9%	0.8%
1995	2.0%	8.8%	1.1%
1996	2.0%	8.9%	1.2%
1997	2.0%	9.0%	1.3%

Percent of adults under correctional supervision by race, 1986-97
Figure 23

In that mostly males populate jails, one can reasonably conclude that in 1997 nearly one out of every five black males were categorized as being either in jail, on probation, or on parole. Furthermore, the trend is clear. Worth noting, the black population in prisons is growing at a faster rate than the white population in prison.

Again, what explains this?

Addressing recidivism, a different aspect of the problem, the Bureau of Justice Statistics reported on July 18, 2004,

> More than half of the nation's 665,475 local jail inmates as of June 30, 2002, were on probation, parole or pre-trial release at the time of their arrest, the Justice Department's Bureau of Justice Statistics (BJS) announced today. Forty-one percent had a current or prior sentence for a violent offense, and almost a quarter (24 percent) had three or more prior incarcerations in jail or prison. More than 60 percent of the jail inmates were members of racial or ethnic minorities, the same as in 1996. An estimated 40 percent of those in jail were black, 19 percent Hispanic of any race, 1 percent American Indian, 1 percent Asian and 3 percent were members of more than one racial or ethnic group. About 36 percent were white.

Reflecting that blacks make-up only about 1/10 of the US population and the above mentioned percentages of blacks and whites in prison, the chance of a black man finding himself in prison is eleven times greater than a white man finding himself in prison. That stark statistic might suggest questioning the processes in our justice system.

With that as background, we now return to the more focused subject of this book. Only partially addressing the issue, blacks in the United States are three times more likely to be disadvantaged than whites. Accordingly, that suggests that blacks are three times as likely as whites to have limited IQs; that blacks might be disproportionately associated with the problems of society. That, however, does not explain the ratio just shown of 11 to 1 for incarcerations. And while the 3 to 1 ratio may be reduced to a 1 to 1 ratio by raising the intelligence of the disadvantaged, and the same for blacks, there are other factors that must explain the remaining dichotomy. There is abundant evidence to show that this disproportionate representation in prison reflects injustices and bias in society, in general, and the judicial system, in particular. Those issues require an additional set of answers and resolutions.

Understanding the causes of crime does not lend itself to simple answers, yet there are clear and related factors. As was just seen, the school dropout rate, as well as the unemployment rate, are closely related to each other, and, they, of course, to poverty. All three are related to intelligence; as intelligence increases, school dropout rate, unemployment rate, and poverty likely decrease. Several types of crime increase with poverty, and also true, several types of crime decrease with affluence. Crime is highly related to intelligence. Sixty-two percent of prison inmates have IQ scores in the lowest 20 percent (that's an IQ of less than about 87).

Given that nearly two-thirds of all prisoners have an IQ in the lowest 20 percentile, one might observe that by increasing intelligence, as measured by IQ tests, over time prison populations would likely decrease. The array of reasons is not obvious, but with increased intelligence, especially among those who had IQs toward the bottom of the spectrum, success and interest in school increases, the likelihood of school completion increases, employment is more likely, and people would be better able to participate in society's wealth and health. They would be busier and more active in contributing to society's well-being, and with that comes more ownership in society and probably increased interest in its well-being, in the improvement of one's own well-being and that of one's family and offspring, and, thus, being less prone to contribute to the problems of society.

The data, more specifically, is described in the line chart of figure 24, using data shown in *The Bell Curve* (p. 376). About 44 percent of the prison population has an IQ in the lowest 10 percent, and another 18 percent have an IQ in the next lowest 10 percent. That is, 62 percent (44 + 18) have an IQ in the lowest 20 percent of the IQ spectrum. This is shown more graphically in the associated pie chart of figure 24.

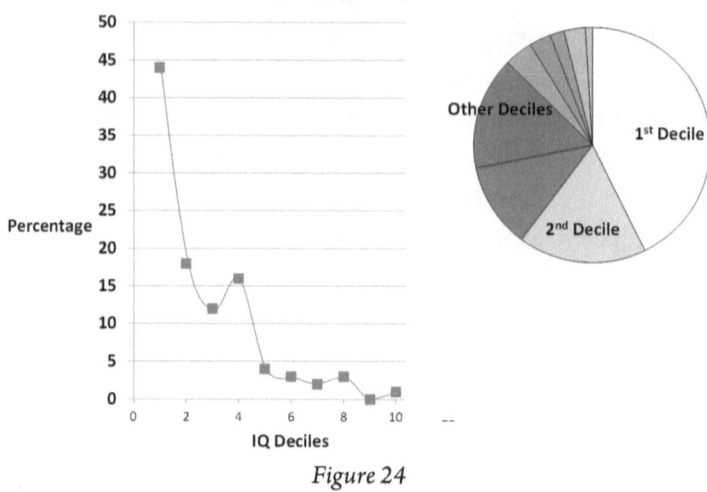

Sixty-two percent of men ever interviewed in prison or jail had IQs in the bottom 20 percent of the IQ spectrum

Figure 24

The observation that intelligence and the problems of society are closely related is not new. Jensen, one of the most respected experts in this field, in 1998, wrote that several hypotheses have been suggested explaining the IQ-delinquency relationship. One hypothesis conveys that having a lower IQ than one's siblings (about ten to twelve points, on average), and classmates, often results in less success and more experiences with failure in school situations. Frustration results, along with alienation and aggression. Inability to settle conflicts verbally often lead to physical threat.

The difficulty with these observations is centered on the comment that these siblings were brought up in the *very same* socioeconomic background. Surely, there were differences, perhaps significant differences. Perhaps one sibling was breast-fed and the other was not. Perhaps one was the first born and the other was not. Surely, the friends of one were different from the other. Perhaps the family experienced difficult times differently with one from the other. It cannot be said what those differences may have been or what caused them, but differences are sure to have been experienced. In any case, it certainly is common and expected for siblings to have a ten-point difference in IQ scores. The one with the lower IQ score is more likely to be involved with criminal activity. So, again, raising the intelligence, at least as measured by IQ tests, reduces the incidence of crime.

Furthermore, Jensen writes in 1998,

> Delinquents and adult criminals typically average ten to twelve IQ points below law-abiding persons living in similar circumstances. Juvenile delinquents who become adult criminals have lower IQs than delinquents who do not become criminals. Recidivists have lower IQs than one-time offenders. ... The peak crime rate occurs in the IQ range of 75 to 90, with the highest rate for violent crime in the range from 80 to 90. (p. 570)

This data supports the apparently simplistic direction to raise intelligence so that crime may be significantly reduced.

Another hypothesis relates to the view that low IQ delinquents have not yet developed the moral reasoning usually reached by adults of similar IQ. Yet other hypotheses relate to impulsiveness, difficulty to delay gratification, and difficulty in seeing the long-term consequences. Jensen wrote, "Whatever its cause, ... the hypothesis that a moderately low level of g [general intelligence][116] is probably one of the most direct causal factors." (p. 297) Thus, one can appropriately conclude that raising intelligence, at least as measured by IQ tests, would reduce many types of criminal behavior.

Substantial evidence appears to suggest that racial biases explain at least some of the discrepancy of black and white incarceration. Specifically, for example, why are black males far more likely to be detained for drug use than white males? Similarly, why are Latinos far more likely to be detained for drug use than whites? Why is a jury far more likely to convict a black defendant for a crime than a white defendant for the same type crime? Why is a jury far more likely to sentence a black man to death for a murder than a white man? There seems to be no doubting that biases prejudice our judicial system.

Bias, a weak word for prejudice, is an integral part of our society. More to the point, essentially each one of us is biased. Please take a little quiet private test. Assuming you knew nothing else about the professional, with a medical

---

[116] In the next chapter, the notion of g is viewed by Jensen and his many followers as one or more genetic factors. As will be seen, this position is shown to be in error; that g is far more likely to be one or more environmental factors.

problem, would you prefer to go to a black doctor or a white doctor? Would you prefer to go to a male doctor or a female doctor? Would you prefer your best friend to be defended for a crime by a black lawyer or a white lawyer? Would you prefer your child's teacher to be black or white? If you were a teacher, would you prefer to deal with black children or white children? If you answered these questions honestly, you probably recognized some biases. By the way, I did not ask you whether you are black or white, male or female. There is no reason to assume blacks are any less biased than whites, or that males are any less biased than females. Essentially, all of us have biases, and, of course, some of us are more biased than others. And some of us are more biased on some subjects than others on the same subjects.

Interestingly, it costs about as much to send a student to Harvard for a year as it costs to support an inmate in prison for a year. The total cost to society for all prisoners is more than 90 billion dollars. Increasing IQ scores to 87 could, over time, reduce prison populations by about 30 percent (see above chart). Even a small increase in IQ scores would have a dramatic impact on prison populations and the related costs to society, not to mention the effect on the people directly and indirectly involved.

Each of the following problems of society will be discussed far more briefly because many of the considerations of intelligence have already been discussed in the previous discussions. Unfortunately, this will also result in brushing aside some of the other factors unique to each of the following issues.

Considerations of children living in poverty

Children live in poverty for many different reasons. Here, we focus on one, the intelligence, more specifically, the IQ of the mother. The information shown in the following chart, figure 25, using data indicated in *The Bell Curve* (p. 383) conveys that the higher the IQ of the mother, the less likely her children will live in poverty. Of course, there are many other factors, but this chart shows IQ and its various implications to be a key factor.

The line chart of figure 25 shows that about 39 percent of the mothers of children living in poverty for the first three years of their lives have an IQ in the lowest decile (an IQ below 81) and about 24 percent of the mothers of children living in poverty for the first three years of their lives have an IQ in the second lowest decile (an IQ between 81 and 87). The accompanying pie chart shows perhaps more graphically that 63 (39 + 24) percent of mothers, whose children live in poverty for the first 3 years of their lives, have IQs in the bottom 20 percent of the IQ spectrum.

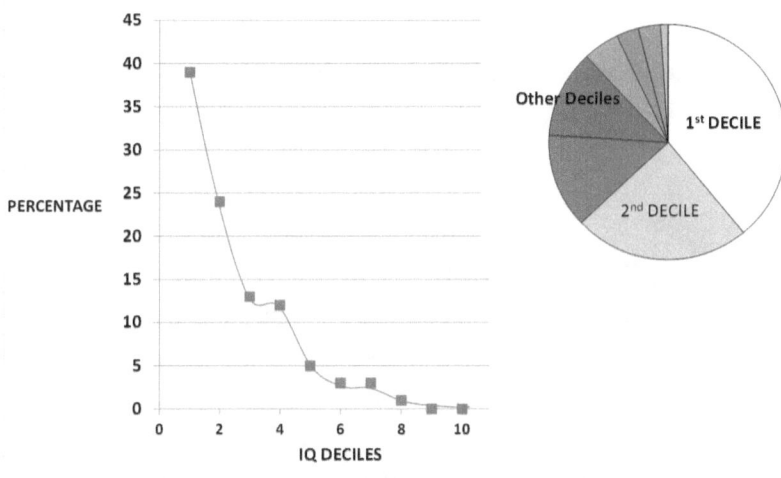

Figure 25

This suggests a rather basic solution—raise the intelligence of disadvantaged mothers, at least as measured by IQ tests, to an IQ of 81. This may reduce the number of children living in poverty significantly. Raising IQ scores another six points, to 87, may reduce the number of children living in poverty significantly more, for a total of about 37 percent, $39 + 24 - 2(13) = 37$. This estimate is derived by assuming that the rate in the first decile might be lowered to the rate in the third decile thus $39 - 13$ or a reduction of 26 percent, and similarly for the second decile with a reduction of 11 percent, $24 - 13 = 11$ with a total reduction of $26 + 11 = 37$ percent.

To clarify this rationale, had these mothers scored higher on IQ tests during their youth, they would have, more likely, completed their education and, more likely, had had opportunities for better jobs. These data show what might have resulted had changes been made early in their lives.

What can be done for them when they reach their teens and twenties? Their intelligence can be raised with nutrition, enhanced health, and more effective educational opportunities. Will their IQs move up to 81 or 87? The answer is yes, but not as easily as could have been done had the enhancements started with their prenatal care. Long-term solutions, starting before birth, are much more fruitful than short-term solutions started later in life. It seems that society tends to focus on short-term solutions because people need quick results. On the other hand, as will be pointed out later again, raising someone's intelligence shows some benefits essentially immediately. Dramatic results are realized over time, measured in months and years, not necessarily decades. Should we pursue the short-term as well as the long-term solutions? Of course!

Chronic welfare

As shown in the following chart, chronic welfare is another problem of society closely associated with intelligence as measured by IQ tests. Specifics, as may be seen by examining the line chart of figure 26, using data shown in *The Bell Curve* (p. 377), indicate that about 34 percent of welfare recipients have IQs in the lowest 10 percent, and another 23 percent in the next lowest 10 percent of IQ; about 57 percent of welfare recipients (34 + 23) have IQs in the lowest twenty percent of IQ, 87 point scores or lower. The accompanying pie chart shows more graphically that 57 percent of welfare recipients have IQs in the bottom 20 percent of the IQ spectrum. And again, the similar message, raise the IQ of the disadvantaged and chronic welfare would be significantly reduced.

The rationale that welfare may be reduced by raising IQ scores follows the same pattern as indicated for earlier discussions of the problems of society, namely, with increased intelligence comes more success in school, and greater likelihood to complete school, thus better job opportunities, more and better employment, and thus less need for welfare.

**Fifty-seven percent of chronic welfare recipients
had IQs in bottom 20 percent of the IQ spectrum**

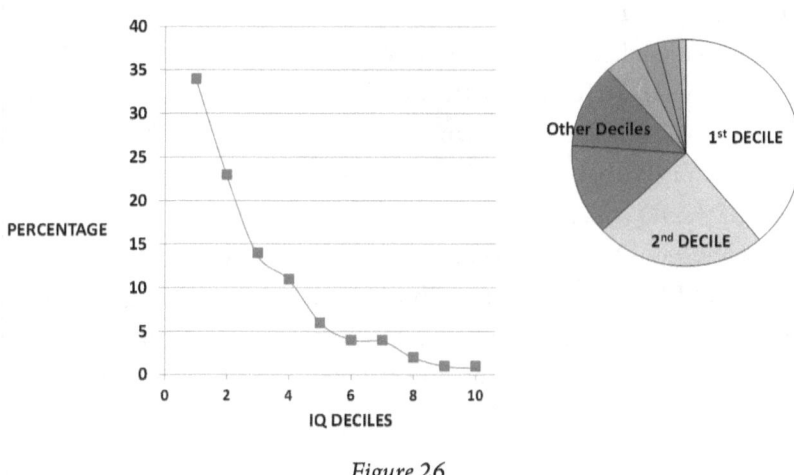

Figure 26

Frequency of low birth weight (LBW) babies

Many LBW babies have health problems and will likely have relatively low intelligence, about thirteen IQ points lower than normal birth weight babies[117]. Blacks have about twice as many LBW babies as whites. One might think this to be genetically caused. To analyze this more closely, UNC psychologists reviewed the data of LBW babies in Africa. They learned that the rate of LBW babies in Africa is the same as the rate of LBW babies for whites in the United States. So, what would explain the dramatic differences in the United States? Could it be explained because blacks have babies on average about one year earlier than whites? Could it be explained because of health considerations? Does the limited prenatal medical care among the disadvantaged explain some of the differences? Perhaps limited nutrition of the disadvantaged explains this.

Of the children I tutored and mentored, one child's mother was thirteen years older than him. Though he was very bright, he likely would have been significantly brighter had his mother been more physically mature.

---

[117] Scarr,

Specifically, 12.9 percent of babies born to black women in 2012 were LBW babies, as compared to 6.9 percent among whites. The ratio was even worse for VLBW babies,[118] being 2.8 percent compared to 1.2 percent.[119] And LBW babies are likely to have not only significantly more health issues but are more likely to have lower IQs.

With that background we can now focus on the frequency of LBW babies according to the IQ level of the mother regardless of race. The following chart, figure 27, using data shown in *The Bell Curve* (p. 381) indicates that 26 percent of LBW babies were born to mothers having IQs in the lowest 10 percent of the IQ spectrum and that an additional 19 percent were born to mothers having IQs in the next lowest 10 percent, for a total of 45 percent (26 + 19) of LBW babies born to mothers in the lowest 20 percent of the IQ spectrum. The accompanying pie chart shows this more graphically. Glancing at the following chart, one quickly realizes that the incidence of LBW babies decreases according to the IQ of the mother.

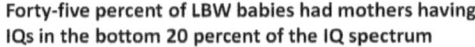

Forty-five percent of LBW babies had mothers having IQs in the bottom 20 percent of the IQ spectrum

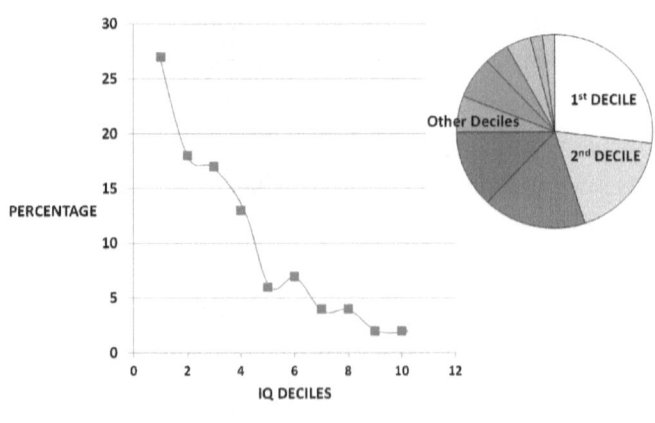

Figure 27

[118] While LBW babies weigh less than about 5.5 pounds, very LBW babies weigh less than 3.1 pounds

[119] Source: Birth Data, National Vital Statistics System http://www.cdc.gov/nchs/births.htm - See more at: http://www.childtrends.org/?indicators=low-and-very-low-birthweight-infants#sthash.GDs3Z0eP.dpuf

This suggests, again, a rather basic solution. Raising the intelligence of mothers, at least as measured by IQ tests, to an IQ of 87 or higher would reduce the number of LBW babies substantially.

This chart does not reflect racial considerations. It merely relates to IQ of the mother. It, therefore, addresses the much higher rate of LBW babies among black mothers who have lower IQs than white mothers, and that because black mothers are much more likely to be disadvantaged than white mothers. The much more frequent occurrence of LBW babies among blacks is explained by the much higher frequency of blacks living in poverty.

The cost of this societal problem may be dramatic. As documented by Almond, Chay, and Lee, in 2005,

> The expected costs of delivery and initial care of a baby weighing 1000 grams at birth can exceed $100,000 (in year 2000 dollars), and the risk of death within one year of birth is over one in five. Even among babies weighing 2000–2100 grams, who have comparatively low mortality rates, an additional pound (454 grams) of weight is still associated with a $10,000 difference in hospital charges for inpatient services. ... Studies have also established a correlation between LBW and high blood pressure, cerebral palsy, deafness, blindness, asthma, and lung disease among children, as well as with IQ test scores, behavioral problems and cognitive development.

Yet again, summarizing as related to previously discussed problems of society, had these mothers scored higher on IQ tests during their youth, they would have, more likely, completed their education and, more likely, had had opportunities for better jobs. They may have delayed having a baby. They may have been better nourished. These data show what might have resulted had changes been made early in their lives.

What can be done for them when they reach their teens and twenties? Their intelligence can be raised with nutrition, enhanced health, and some educational opportunities. Will their IQs move up to 87? Perhaps the answer is yes, but not as easily as could have been done had the enhancements started with their prenatal care. Surely, better nutrition, alone, during pregnancy would reduce the likelihood of having a LBW baby.

Illegitimate children

Single mothers having babies occurs far more often among all races and all socio-economic groups today than in the later part of the twentieth century. Many reasons explain this. We do not attempt to explain this recent development in the United States and elsewhere in the world. Suffice it to say that much of this increase perhaps is due to women's choices.

Here, we discuss only one aspect of this phenomenon that was far more prevalent in the past and remains an explanation for much but definitely a decreasing percentage of illegitimate children. As shown in the following chart, figure 28, using data shown in *The Bell Curve* (p. 378) indicating data prevalent about two decades ago, the greater the intelligence of the mother, the fewer illegitimate babies.

First, please notice the graph is not nearly as skewed as other such charts shown earlier in this chapter. Even the data for decades ago indicate that women of higher intelligence chose to have babies out of wedlock, but in decreasing frequency according to intelligence. This graph today, if the data were to be collected, would indicate more mothers with greater intelligence are having children out of wedlock. The point of this section is to explain not the entire spectrum shown in this graph, but only the greater frequency of illegitimate children born to women of limited intelligence.

Fifty-two percent of illegitimate children were born to mothers having IQS in the bottom of the IQ spectrum

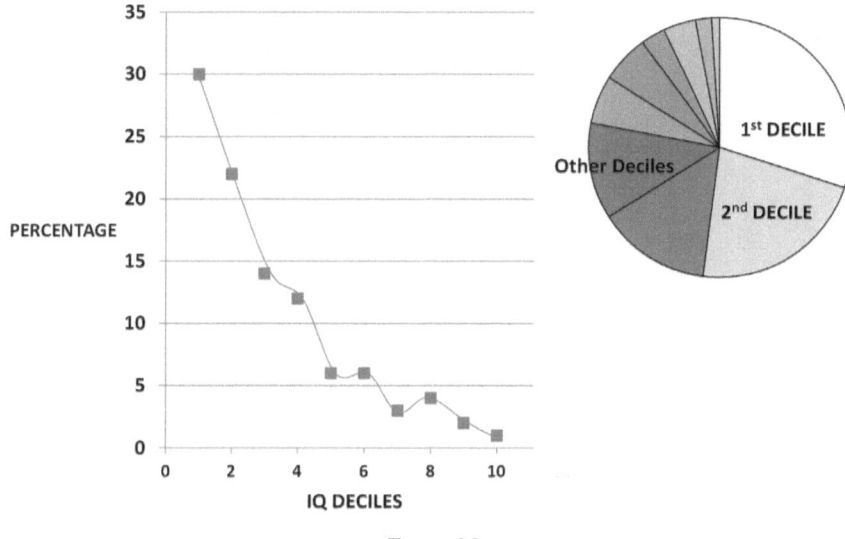

Figure 28

This chart suggests, yet again, a rather fundamental solution—raising the intelligence of mothers, at least as measured by IQ tests, to an IQ of 87, would reduce the number of illegitimate children significantly.

Again, with apparent cultural changes in recent decades, intelligence likely has less relevance with this societal and cultural aspect than appeared to be true twenty years ago.

Automobile accident and fatality rate

A myriad of other social problems may be similarly reduced by raising intelligence, especially of those toward the low end of the spectrum. For example,

A longitudinal study of Australian servicemen, none of them in the [mentally] retarded range, found that IQ correlated with rate of death by automobile accident, even after controlling for other characteristics (O'Toole, 1990). The auto fatality rate for men of IQ 85 to 100 (92.2 per 100,000) was nearly double that for men of IQ 110 to 115 (51.5 per 100,000). The rate for men of IQ 80 to 85 was about three times as high (146.7 per 100,000).

This is shown graphically in the following figure, figure 29.

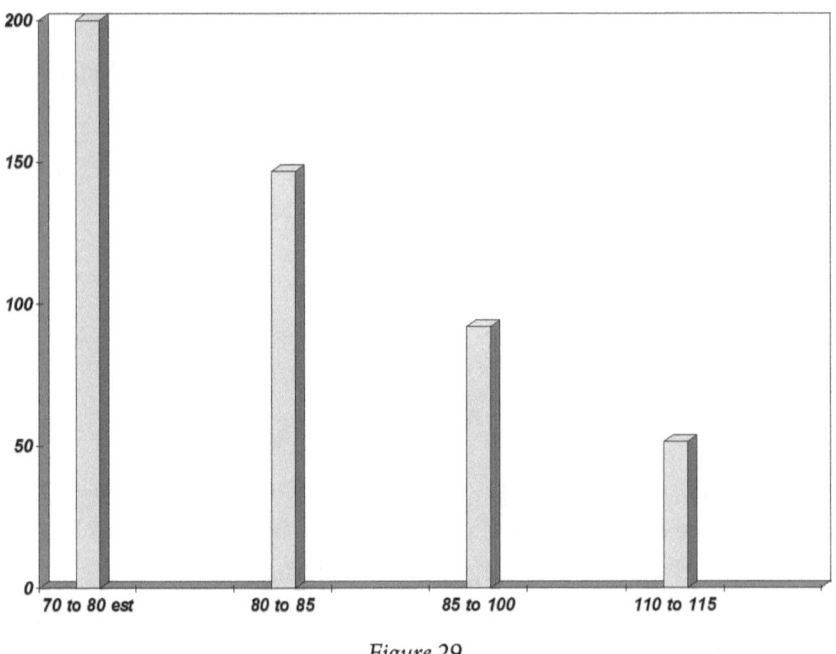

Figure 29

Surely, a wide array of circumstances may explain these data, but the stark contrast of IQ differences with death by automobile accident seems to be additional evidence to enhance IQs.

Other problems include,

- Medical problems are often prevented or limited when following medical instructions either from a pharmacist or a doctor. The more intelligent a person, the more closely medical directions are followed, and the less frequent and less extensive the medical problems.

- Frequency of accidents resulting in disabilities in the work place, assuming the same types of work, is closely associated with intelligence. Those with greater intelligence tend to have not only fewer accidents but also less severe accidents.
- Absenteeism in the work force is associated with well-being which is related to intelligence.
- Costs for medical care are several times higher for the disadvantaged than the advantaged

Reviewing these individually one finds support for the notion that by increasing intelligence, especially and most readily of the disadvantaged, one significantly reduces the problems of society. When reviewing them collectively one is increasingly assured of this outcome.

Summary

Repeating a chart shown at the beginning of this chapter summarizes much of the information described in this chapter.

As may be seen immediately upon reviewing figure 30 most of the problems of society are associated with low intelligence; the problems of society are disproportionately associated with those scoring in the lowest 10 percent of IQ scores with geometrically less involvement with each increasing 10 percent of IQ scores.

This graphic conveys the message powerfully: increase intelligence and the problems of society are likely significantly reduced. Basically, raising the IQ of those in the lowest 10 percent into the second ten percent may reduce that part of the societal problems by 20 to 25 percent; raising the IQ of those further into the third ten percentile may reduce that part of the societal problems by another about 12 to 15 percent.

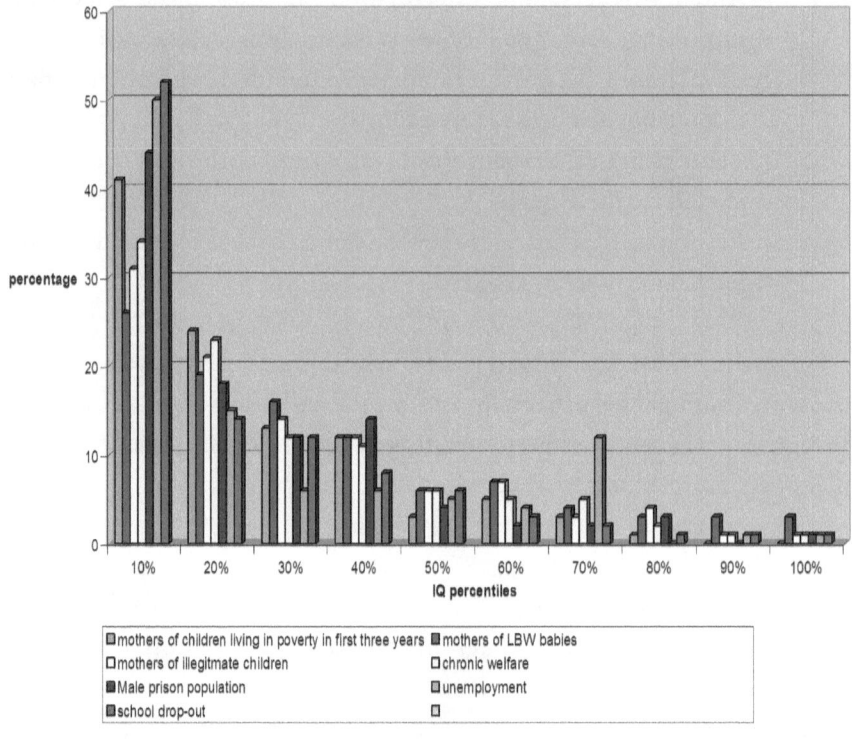

Summary chart of relationship between IQ and problems of society
Figure 30

More specifically, and realistically, if intelligence of only the disadvantaged were to be increased by an average of about eight points, most of the lowest 10 percent would have IQs above 81, the start of the second 10 percentile. By how much would the school dropout rate decrease? The answer is perhaps by as much as nearly 20 percent. Accordingly, and given a stable society with a reasonably fully employed workforce (3 to 5 percent unemployment), how much would the unemployment rate drop? Perhaps by about 30 percent. By how much would the prison population decrease over time? Given the school dropout rate reduction and the unemployment reduction, the prison population may drop, perhaps, by as much as nearly 30 percent. Undoubtedly, some will feel these observations to be too optimistic. Maybe so! Regardless, rationally, one would likely conclude, because of the reduction of these problems in society, the advantages to society would be significant. And this list of advantages only begins to tell the story.

Of course, there are other factors to be considered, but raising intelligence seems to be one of the most central ways to resolve these issues.

More on causal relationship

As asked earlier, does low intelligence *cause* poor health, significant school dropout rate, and persistent joblessness? With the data just shown in this chapter, this basic question may now be more readily answered. The data, for each of these areas of concern, are associative; the data itself does not directly indicate that low intelligence causes each of the problems of society. However, when examining unemployment, school dropout rate, or any one of the other areas discussed in this chapter, causation is increasingly strongly suggested; the patterns are very clear. More importantly, the implication gets stronger with each additional area of concern examined.

Asking a related and differently difficult question, why would, how could, a correlation be viewed as causal? Causation cannot be concluded directly from a correlation. Whenever there is a correlation between two characteristics, for example between IQ scores and school dropout rate, or more generally, between IQ test scores and each of the various problems of society, analysis is applied to understand, to make sense of, the correlation. While causation cannot be surmised from the correlation data itself, via close analysis several clear possibilities surface. Either characteristic A, say limited intelligence, causes or affects characteristic B, say school dropout rate, or the other way around, B affects or causes A, or one or more other underlying factors cause or affect both characteristic A and characteristic B. The challenge is to rationally explain and justify the causes and the direction of causation. Furthermore, if there is only one association, say, under-employability and IQ, causation may be unclear, but with each additional association, school dropout rate and IQ, in the same direction, causation becomes increasingly likely.

Rational examination of the known underlying relationships in each of these areas discussed helps one clarify whether, and to what degree, the data, while associative at the surface, proves to be also causative. We know that an undernourished, or an even just somewhat undernourished individual, at any given moment, behaves less intelligently than when well-nourished. We know that over time such an individual gains, retains, less information from experiences than one who is well nourished. We know that over time intelligence, as measured by IQ tests, gradually and accumulatively drops for those that are less than well nourished. We know that a person with lower IQ will more likely fall behind in school, be bored by the material, gradually lose interest and motivation to stay in school, gradually feel less than able to succeed in school, and tend to drop out of school. A school dropout is less employable, and even when employed will, more likely, receive poor wages, be disadvantaged, and provide less adequately for the members of his or her family. The children will likely be less than well nourished, and the cycle repeats itself for the next generation, and then the next generation. So, it is reasonable to conclude that a school dropout is less employable; that's causal. We know that someone who is less employable is more likely to live in poverty and receive welfare; we know these associations are causal.

While one cannot claim the data to be conclusively causative, the more supportive the rationale, the more one may rationally conclude the associative information to also be causative. The more closely one examines the source of smoke, the more apparent the cause may prove to be fire.

If the answer to these questions is that the relationship between low IQ scores and the problems of society is causal, then it is clear what to do. Surely, limited intelligence inhibits the effectiveness of one's educational opportunities and contributes to school dropout rate. A combination of low intelligence and less than effective educational experiences limit job opportunities which inhibit economic security, and often limits social status. All of these seem to relate to motivation, self-confidence, and success (whatever that may mean). And at least one underlying explanation for these collective concerns is limited intelligence. Therefore, to improve these characteristics, one should improve health and nutrition, and thus raise the effectiveness of formal and informal learning experiences and thereby, indirectly, increase intelligence; something we know how to do.

People with limited intelligence, whether intelligence is measured by IQ scores or not, likely make less than effective decisions. This is particularly important regarding the issues causing high school students to drop out of school, relating to criminal activity, unemployment, chronic welfare, having LBW babies, having serious automobile accidents, and other problems of society.

Also, the more likely one lives in poverty, the worse the level of nutrition, the more likely to have poor health, the more likely one has difficulty on the job, and the more likely to be unemployed. Of course, an implied aside is, the more likely one lives in poverty, the poorer one's health, and the greater the potential need for health care.

None of these observations are absolute, but all of them rationally help explain contribution to specific problems of society, and each of these can be improved by raising the level of intelligence, especially of the disadvantaged.

Other areas

The next chapter focuses on several stubborn and basic myths that have plagued the study of intelligence for many generations. These myths need to be exposed to the light of day and placed into the annals of psychology's archives so that research may proceed unencumbered and so that progress may be more effectively achieved.

Because intelligence, as measured by IQ tests, is substantially determined by environmental factors, at least among the disadvantaged, observations about intelligence are significantly, if not largely, about environmental conditions. This comment has far reaching effects. The Flynn effect, for example, results from the observation that IQ scores have been increasing across the entire planet at about the rate of three points per decade. Significantly, three IQ points, is about 10 percent of the difference between being labeled mentally retarded and viewed as average in intelligence and has far reaching implications. Of course, this observation results from the examination of intelligence tests, IQ tests, which are behavioral tests, and thus influenced by both genetic factors and the environment. Because the tests are adjusted about every decade, to reflect the Flynn effect, this observation, as will be seen, has far-reaching implications on the labeling process, and especially on those labeled mentally retarded and learning-disabled.

The following chapter will address the periodic adjustment of tests and other aspects of IQ testing that have erroneously come to be assumed valid. While IQ tests were originally designed, about a century ago, to predict potential in an academic environment, they were not intended to predict future intelligence. Binet, the developer of the first intelligence test, was strongly opposed to viewing his tests as predicting a stable potential intelligence. He knew intelligence is environmentally modifiable. Furthermore,

IQ tests were never based on scientific study. They were never intended for the purposes their subsequent supporters pursued. This statement may be disturbing to many readers and especially disturbing to educators. Details of this observation will be conveyed in the next chapter.

Additionally, while IQ scores may predict the potential of groups of individuals, they cannot predict the potential of any one individual. As mentioned just above, the Flynn effect, shown to result essentially from environmental improvement, erroneously results in periodic renormalization of IQ tests. If IQ tests continue to be renormalized, then, over time, the average IQ score will be that of Einstein, and, yet, 2.5 percent of the population will still be labeled mentally retarded. Interestingly, that would classify people with IQ scores like current graduate-school students as being mentally retarded. Absurd! The IQ tests might rather be left to float, much like the metrics of height, and the average, currently set at 100, be gradually and naturally raised while allowing the definition of mental retardation to remain at 70. Over time the number of mentally retarded individuals will thus diminish, as appropriate, as IQ scores increase.

These and other areas of concern regarding the understanding and interpretation of intelligence are discussed in the next chapter.

CHAPTER 7:

The Mythology, Magic, and Mystery of Intelligence

Myths of intelligence delay understanding intelligence

Each chapter of this book has contributed differently to the tapestry of this book. This chapter introduces threads that provide the tapestry with strength and cohesion. Some threads described in this chapter reveal myths that have blocked the path, delayed progress, to understanding intelligence. Some threads describe delayed progress to solving basic and widespread human problems. Other parts of the tapestry would be complete without these threads, but these threads are needed to prevent the tapestry from unraveling.

Essentially every field inadvertently develops myths which need to be recognized as such, discredited, and then securely placed into the field's mythology. Medical professionals know of many myths that have plagued development of the medical profession for centuries; the value of blood-letting is one of those myths. Psychologists know of many myths as well, including the once serious study of phrenology; the study of bumps on the skull.

In the study of intelligence, myths have developed that have spawned other myths, which have in-turn created yet others. These have not only impeded progress but have severely hurt a great number of people, billions of people throughout the world, for generations. Some myths result from poor analysis, some from poor research processes, and some from well-meaning people reacting to a narrow set of experiences. All professionals, I'll venture to claim, all people, have filters, biases, through which they study and recognize phenomena. Honestly derived notions sometimes become accepted

221

myths, which over time become self-perpetuating, and, in fact, become the foundation for a pyramid of myths.

A word of clarification is essential here. The word *myth* might seem too strong to describe notions that have proven, at best, misleading in the study of intelligence. To protect against this possible criticism, I will qualify the use of the word where it needs softening.

The understanding of human intelligence has been severely delayed because many myths, though repeatedly questioned, have not been adequately refuted, debunked. So far, this book has touched on several myths, one being the notion that intelligence is largely determined by genetic factors. The evidence that this is a myth was partly demonstrated by Turkheimer and associates. It was discussed in Chapter 4. In truth, while the difference in intelligence between hamsters and humans is surely largely genetically determined, *intelligence of humans, especially the disadvantaged, is largely environmentally determined.* (Possibly the case may be made that differences in intelligence among hamsters is also largely environmentally determined. It seems rational.)

As Turkheimer and associates have shown, the intelligence of disadvantaged children is largely environmentally determined, in a ratio of 20 to 80; that is, 20 percent genetic influence and 80 percent environmental influence, or even as low as 10 to 90. This is in direct contrast with the findings of Jensen as well as those reported by Herrnstein and Murray, who claim intelligence is largely determined by genetic factors. Jensen reports the ratio to be 80:20 – the inverse of that reported by Turkheimer and associates. The reason for the difference is, again (and I apologize for repeating), that while Turkheimer's sample consisted of disadvantaged children, children having had limited nutrition and health, as well as limits in effective formal and informal learning experiences, Jensen as well as the Herrnstein and Murray referenced studies that assumed subjects had average well-being. Jensen's studies appeared to have subsumed the effect of adequate nutrition and good health. When these environmental factors are subsumed, their effect is hardly acknowledged as part of the environmental influence, though they certainly have a strong influence. Thus, though the findings of those whose research is based on individuals with average well-being appear valid (finding that genetic factors have a relatively large influence on intelligence), when factoring in the influence of their already strong environmental influences, the overall influence of environmental factors increases while influence of genetic factors decreases.

Recalling from Chapter 4, let's assume Jensen's ratio of 80:20, that genetic influence on intelligence is 80 percent. Now let's assume the validity of several studies reported earlier in this book concerning just nutrition and well-being; for example, that providing mineral and vitamin supplements to a disadvantaged pregnant woman increases the IQ of her baby by 8 points and providing mother's milk to her baby improves the baby's IQ by 7 points, improving IQ by an average of 15 points (8 + 7). Now let's assume, here, however, that instead of considering a disadvantaged woman and her child, we consider an advantaged woman and her child. Let's assume the advantaged women is less than well-nourished during pregnancy, and that formula is fed her child rather than mother's milk. We would likely find that this child's IQ would be about 15 points lower than had he or she been prenatally and perinatally nourished as would a typical advantaged child. If increasing a *disadvantaged* child's IQ 15 points by providing both mineral and vitamin supplements prenatally and providing mother's milk instead of formula perinatally, then an *advantaged* child's IQ would likely be decreased by about the same amount if the mother were inadequately nourished prenatally and if her baby were provided formula instead of mother's milk. In general, it is likely that an advantaged child, when experiencing the same environmental constraints as a disadvantaged child, will have a similar IQ as a disadvantaged child; that Jensen's ratio of 80:20 will be realized as Turkheimer's ratio of 20:80.

This notion is so central it demands additional clarification. As stated, both genetic and environmental factors determine intelligence. Of that there is no dispute. Furthermore, environmental factors significantly determine the expression of genetic factors. Of this, too, there is no doubt. Thus, crucially, while limited well-being, typical among the economically disadvantaged, constrains the expression of genetic factors, stronger well-being, typical among the advantaged, supports a stronger expression of genetic factors. Additionally, the more adequate the nutrition the more effective the neuronal system. This is likely true for all stages in life. Notably, the test results described previously, using individuals with average socio-economic-status, subsume the environmental influence that permits cognitive genetic factor(s) to be more extensively expressed and the neuronal system to perform more effectively. Therefore, those tests do not reveal an appropriate ratio of genetic and environmental influence on intelligence. In contrast, the tests using individuals with low socio-economic-status do reflect the environmental constraints common among the disadvantaged that influence intelligence by limiting the expression of genetic factors and constraining effectiveness of the neuronal system; genetic influence appears less pronounced. In both cases, environmental factors likely influence intelligence *identically*, the advantaged positively and the disadvantaged negatively; one only appears much more significant than the other. This addresses only intelligence in the moment.

More effective expression of genetic factors allows more effective intelligent behavior in the moment and, thus, also more effective accumulation and integration of new information used for subsequent intelligent behavior. This reflects not only intelligence in the moment but also intelligence growth.

Stated crisply, if the IQ of a disadvantaged individual were *raised* by reducing specific environmental constraints, then burdening an advantaged individual by those same environmental constraints would likely cause his or her IQ to be *lowered* by about the same amount.

It is therefore concluded that differences in intelligence among humans are largely determined by the environment. Surely, some individuals at the extremes are more influenced by genetic factors but for the clear majority, *differences in intelligence among humans are largely determined by the environment perhaps in the ratio of 20:80, or even lower, as Turkheimer and associates have shown.*

Yet another myth, also discussed earlier, is the myth that differences in intelligence between groups of individuals are determined, at least partly, by genetic factors. Some say that the genetic material differ between advantaged and disadvantaged peoples, and also between blacks and whites, such that intelligence differs when influenced by the same environmental factors. This does not hold under scrutiny; currently, there is no evidence to support the notion that genetic factors explain any differences in intelligence between groups of individuals. If such information were revealed then, of course, this strong observation would require review, reanalysis, and perhaps retraction. In any case, environmental influence has been shown to explain the differences in intelligence between groups of humans. I hasten to add, differences between a group of Nobel Prize winners and most any other group of people may be shown, someday, to be partly genetically determined. The focus here is differences between racial groups as well as disadvantaged groups of people and advantaged groups of people, and these differences have been shown to be *entirely* environmentally, not genetically, determined.

The myth that intelligence is stable was also discussed earlier. It was shown that IQ scores, being age-independent, are stable by design, whereas *intelligence, being age-dependent, growing in each developmental stage, cannot be stable.*

This book, through quantitative data, showed these myths to be just that, myths. Recognizing these myths, as such, should finally permit the understanding of intelligence to grow and expand unobstructed. Because intelligence is central to the human condition and, of course, being a major part of the field of psychology, it affects many other fields such as education and sociology. To achieve an understanding of intelligence requires the myths to be revealed, disputed, discredited, and discarded; to become respected and vital, the study of intelligence must be unburdened of its myths.

These myths have been with us for generations, and some for centuries. Some of the most respected researchers, several mentioned in this book, have either created these myths or strongly supported them. Injustices caused by myths cannot be corrected; they can only be recognized and, at best, prevented from affecting people and research in the future.

Of course, we will mention whether and where a myth may be controversial, disputed, but otherwise will merely call a myth a myth.

By extension, Chapters 4 and 5 provided evidence to help expose another set of myths. The view that the disadvantaged are innately limited in intelligent behavior can now be definitively labeled a myth, and by extension, it is a myth that blacks are innately limited in intelligent behavior— myths that have hurt tens of millions of people in the United States and billions around the world. Reducing environmental constraints of a disadvantaged person will likely significantly raise his or her intelligent behavior depending on his or her level of disadvantage. Currently, blacks in the United States are far more likely to be disadvantaged than are whites, and, blacks that are disadvantaged are likely to be more disadvantaged than whites who are disadvantaged. As shown earlier in this book, the characteristics of disadvantage, alone, explain the entire IQ gap between blacks and whites. This is even more evident in other parts of the world, such as sub-Saharan Africa.

The emotional notion that "I am black; I am stupid" as children have told me, or "I am a girl, so I can't do math" can now be recognized as biases and its past consequences need not reoccur.

In net, the IQ gaps between the advantaged and disadvantaged, the blacks and whites, as well as probably the Asians and whites, are explained by well-being, motivation, expectations, and other environmental differences.

While several attempts have been made to present arguments to the contrary, they have been repeatedly rejected,[120] yet some researchers will perhaps continue to try to show otherwise. Knowing this, and having a need to move on, this observation is somewhat softened here. We maintain that differences in intelligence between groups of humans are greatly determined by environmental differences experienced by the groups. This set of words is not in dispute and will be the focus of the next few sections. Surely, some genetic differences in intelligence between groups may surface in the future; however, perhaps they may only show differences—not superiority. For example, an Olympic winner in breast-stroke is different from an Olympic winner in the back-stroke, but one is not necessarily better than the other.

Thus, the following previously discussed myths may be crisply identified,

1.   *It is a myth that differences in intelligence among individuals are largely determined by genetic factors.*

    Evidence shows that differences in intelligence among individuals is largely determined by environmental factors.

2.   *It is a myth that differences in intelligence between groups of individuals are even partly genetically determined.*

    To the contrary, differences in intelligence between groups of individuals are entirely environmentally determined; and this should hold until evidence surfaces showing otherwise.

Accordingly,

3.   *It is a myth that the disadvantaged are genetically more limited in intelligent behavior than the advantaged*

---

[120]   See exchange of articles between Rushton and Jensen, and, Sternberg in 2005

And

4. *It is a myth that blacks are genetically more limited in intelligent behavior than whites*

Furthermore,

5. *It is a myth that intelligence is stable over time.*
    To the contrary, while IQ is defined, for convenience, to be stable over time, intelligence of individuals typically grows over time, and thus, is not stable.

Armed with this awareness, we can now question several additional stubborn presumptions and myths that have plagued the understanding of intelligence by professionals and laypeople for many decades; one can proceed to understand the more general implications of the myths already mentioned, and others as we proceed.

We know that the fruits of the environment are not uniformly distributed, witness the fact that about eighty percent of the world's population lives on less than \$10 per day, nearly half the world's population lives on less than \$3 per day; over a third of the people in the world lives on less than \$2 a day (See Figure 2). Given that intelligence is largely determined by the environment, it seems clear that we can do much to improve intelligent behavior of most people in the world and of nearly every child in the world, even our own.

Several additional stubborn myths may now be identified as such. Given that the fruits of the environment are not uniformly distributed, and that these environmental factors influence many types of human behavior not only intelligent behavior but also height, we now may question the belief that intelligence-measures should present a normal bell curve distribution. In particular, one may question why IQ test scores should present a distribution with a mean of 100, and with each of the positive and negative standard deviations equal distances from the mean; the first standard deviation (SD) being, say, + and - 15 points from the mean, the second SD being, say, another + and - 15 points, that is, + and – 30 points from the mean, the third SD being another + and – 15 points, that is, + and – 45 points from the mean, and so forth. Specifically, how can intelligence-measures present a bell curve distribution when the environmental advantages, that significantly influence intelligence are not uniformly distributed? And by extension, this belief, that intelligence-measures present a bell curve distribution, has naturally led to the impression that IQ tests should present scores forming a bell curve distribution. If intelligence-measures cannot, should not, form a bell curve distribution, IQ tests should not be defined to yield scores[121] forming a bell curve distribution either, and they should not be periodically renormalized. Yet currently, because IQ scores are increasing at about the rate of 3 points per decade, IQ tests are renormalized every ten to twelve years assuring the persistence of the notion that IQ scores present a normal bell curve distribution, and, encouraging the belief that intelligence-measures present a normal bell curve distribution.

Expanding on these concerns, why do we allow children, our children, and adults to be labeled according to their IQ scores that may be changed merely by enhancing their environmental conditions such as nutrition? Children labeled as learning-disabled have been prevented from receiving services unless their IQ is higher than average. Why do we inhibit children with real learning disabilities according to their IQ scores? Why do we label individuals as mentally retarded when we know that given better nutrition their IQ would very likely move into the near average range; when, if these same people were given the IQ tests used several decades ago would have been considered near average? Should the practice of punishing a convicted murderer be dependent on an IQ determined several decades ago? Should someone with an IQ of 70 be prevented from receiving the death sentence, perhaps given life in prison, whereas someone with an IQ of 71 perhaps given the death sentence? Surely, we know that this one-point difference in IQ could have been caused by differences in momentary cognitive genetic factor influences, but we know it is more likely to have been caused by momentary difference in environmental factors.

Knowing that IQ scores have increased steadily for many decades,

---

[121] By way of clarification, this myth regarding IQ testing has resulted in the requirement that all IQ tests be defined to yield scores that average 100, with 1, 2, 3, and more SDs from the mean being precisely defined; that, for example, all individuals with an IQ at or below -2 SDs (70) be labeled mentally retarded, regardless of the upward movement of intelligence across the entire population.

shouldn't we change the various categorizations of intelligence such as mental retardation? In particular, given that intelligence is increasing, especially among those toward the low end of the intelligence spectrum, should we continue to require that about 2 percent of the population be categorized as mentally retarded? Shouldn't we allow that percentage of people categorized as mentally retarded to decrease as people at the low end of the spectrum become more intelligent? Certainly, we should recognize that the number of people in this and other categories is steadily changing. Given the erroneous categorizations of intelligence, and the inappropriateness of labeling people in general, shouldn't we reconsider the many legal labels we place on people such as learning-disabled?

And these are just a few of the practices that need to be questioned. They are based on what will be shown to be myths. Other practices, also to be addressed, pertain to research presumptions that have severely hampered our comprehension of intelligence.

These questions are important to each one of us because the perpetuation of myths about intelligence have wide implications on every aspect of our lives. They severely impact much of psychology, sociology, education, politics, to mention just a few of the disciplines affected. They affect all in society, and sometimes can be devastating to our potential and that of those near us, our self-satisfaction and of those near us, our happiness, and, of course, that of our children. Partly because of these myths, much of human potential, especially among the disadvantaged, is, hardly recognized, and, hardly tapped. That is not only true in this and other developed countries, but it is more profoundly true in essentially every developing country in the world.

Many researchers do not know or are too cautious to pursue the broader implications of results perhaps because of concern to question ingrained notions which over time have become assumptions. Without definitive results professionals generally refrain or hesitate to pursue implications that are contrary to accepted positions, sometimes myths. Here, however, the data appears overwhelming strong. Questioning is not only appropriate. It is required.

After conveying, with extensive data in Chapters 4 and 5, that intelligence is significantly influenced by nutrition, health, experiences and other environmental factors we addressed, in Chapter 6 we discussed the implications of that on the problems of society, prison populations, school dropout rate, unemployment, and much more. In this chapter we take this an additional important step recognizing first that intelligence of *most* people in the world is not largely determined by genetic factors but rather largely determined by environmental factors, and then addressing the broader implications of this on society, on education, on psychometrics, and on legal issues. The implications are many. They are presented in an order that essentially builds a growing foundation, each successive building block supported by underlying blocks.

To be precise, we are not really addressing cognitive genetic *potential* here, but rather the potential of one's intelligent *behavior*; human behavior is influenced by the environments previously encountered. This is not only true for intelligence, but it is also true for other human behavioral characteristics.

This chapter deals with several notions basic to an understanding of human intelligence. The discussion does not directly address the central message discussed so far in this book of raising human intelligent behavior, especially of the disadvantaged, the world over, so that poverty is greatly reduced; that economic inequalities and opportunity inequalities may be significantly reduced; that under-employability may be significantly reduced; that learning, formal or informal, may be made more effective; that the human potential, of every one of us, may be more closely reached; and such that the problems of society may be substantially reduced. I have included this chapter to provide a more complete view of the issues that contribute to a better understanding of the interrelated intelligence issues before going on to the last and concluding chapter. Each of the following sections in this chapter separately deals with specific topics, additional myths of intelligence, in turn. Their interaction and their collective messages convey an overall concept of intelligence with a foundation different from that which is widely held today. A more encompassing and comprehensive paradigm of intelligence surfaces.

Intelligence and its measurement affect essentially every field of human activity; the extent that these comprehensions of intelligence are wrong determines injustices and mistreatment of tens of millions of people in the United States alone.

We now address specific notions of intelligence that have misdirected our focus and delayed progress towards understanding intelligence. Some of these have become so ingrained, and so deeply presumed, that they appear inappropriate of questioning or discussion. For example, as mentioned, how can intelligence present a bell curve distribution knowing that intelligence is substantially influenced by the environment and knowing that the environmental advantages are not uniformly distributed? And given that question, how can we restrict some children from gifted and talented learning groups that are based somewhat on IQ scores, when we know these scores can be modified with changes in nutrition or other environmental factors? What are the implications if these current basic notions are incorrect? How do they affect our educational processes, our testing assumptions, the labeling of children and adults, our social behavior towards ourselves and others, and how does it affect our legal code? Are there social injustices that result from misunderstanding these and many other notions of intelligence?

While not directly relevant, I am driven to convey an often-related problem in the medical research field. Before Ruth Kirschstein questioned the practice, drug tests were done on men, not women. This appeared to be a matter of convenience; men do not become inconveniently pregnant, nor do they have menstrual cycles. Kirschstein questioned whether the tests, therefore, adequately surface medical concerns regarding women. I won't go into details, but the issue had and seems to continue to have broad implications. Likewise, determining the relative influence of genetic and environmental factors on people using only individuals with average socio-economic-status is and has been misleading. Results, accordingly, that intelligence appears largely influenced by genetic factors have led to severe delays in understanding the most central facts of the human condition; has led to persistent myths that need to be exposed, demystify, and discredited.

Myth that human characteristics, in general, present a normal bell curve distribution

We assume intelligence-measures present a normal bell curve distribution, but why? The answer lies partly in the widely held belief that other human characteristics such as height appear to present a bell curve distribution. However, are these other generally held beliefs valid? We know, actually only feel, that measured height of a large population, of men or women, presents a bell curve distribution. But height, like, perhaps, other human traits, is significantly dependent on environmental factors such as nutrition. The poor, or more specifically, the undernourished, on average, do not reach the same heights as the well-nourished. For example, North Koreans are significantly shorter than South Koreans. There is abundant reason to believe that that is caused by the current undernourishment of North Koreans.

That was not always true. One hundred years ago, before political differences split the Korean peninsula, Korea was a single nation and a single people. The range of heights for the populations, for men, for example, in the northern part and southern part of Korea was likely very similar. Heights of the people, at the time, *might* have presented a normal bell curve distribution, but not likely; then and now, in most of the world, the disadvantaged are much more prevalent than the advantaged – their limited nutrition cause the curve to be skewed. Likewise, while the height of North Koreans *might* present a bell curve distribution, and while the same may be true for South Koreans, were these populations to be pooled today they would demonstrate a bimodal distribution, with two humps, like a Bactrian camel—not a normal curve distribution. The reason is probably not genetic because their genetic pools have only been separated for several decades. So, the non-normal distribution of scores of heights across North and South Korea must be caused by environmental factors. If environmental factors were uniformly distributed the scores of height would more likely present a normal distribution. But within the Koreas, scores of height cannot be normally distributed because environmental benefits are not uniformly distributed, and of course this is true for essentially every part of the world.

To expand on this, it was documented several years ago that the average "male height in impoverished Vietnam and North Korea remains comparatively small at 5ft 4in (1.63 m) and 5ft 5in (1.65 m) respectively." Currently, young North Korean males are significantly shorter. This contrasts greatly with the extreme growth in height occurring in surrounding Asian populations with correlated increasing standards of living. Young South Koreans are about 3 inches (8 cm) taller than their North Korean counterparts, on average. This is clear evidence that height does not present a normal bell curve distribution; that height while, surely, partly genetically determined, is significantly environmentally determined; any characteristic that is environmentally influenced cannot be normally distributed. Not only are environmental factors not uniformly distributed, but there is no reason to conclude that even genetic influences are normally distributed.

Loehlin and associates report,

> There is an extensive literature on the increase in average stature during the last 150 years in the United States, Western Europe, and Japan. The data show that there has been a general increase of about 1 cm per decade (about 1 inch per generation) over the last five or six generations (Genoves, 1970). ... Kimura (1967) ... estimated that the Japanese population at age 20 has increased by about 6 cm among males and about 5 cm among females during the first 50 years of [the 20th] century. (p. 93)

Earlier findings support this as well. Bloom writes,

> There is considerable evidence, summarized by Sanders (1934), to demonstrate that at maturity the average difference in height between high and low socio-economic groups range from 8 to 20 centimeters, or an average of about 5% of mature height. (p. 37)
>
> [Bloom continues conveying that] Boas (1911) obtained height measurements on children born in Europe and in the United States of the *same* parents. He found ... that children born in this country [the United States] after their parents had been here four years or more were considerably taller than their siblings born abroad or their siblings born in this country shortly after the parents had migrated. The differences were of the order of 5%. (p. 37-38)

Bloom latter indicates "It is likely that the environment could affect the development of stature more during the period of birth to 3 years than during the age period 3 to 6 years." (p. 211)

Collectively, these studies, and others indicated earlier in this book, provide evidence that height is significantly environmentally influenced, and therefore does not present a bell curve distribution. There is abundant reason to conclude from this that other human characteristics significantly influenced by the environment, similarly, cannot present a normal bell curve distribution. Interestingly, the measurement of height is not forced to present a bell curve distribution; scores of height are permitted to float, as appropriate.

Now let's focus on human intelligence.

Is human intelligence normally distributed?

If the environment were the same for all individuals, would intelligent behavior measures present a bell curve distribution? We really cannot say for sure because we do not know that genetically determined cognitive abilities present a normal bell curve distribution. Nevertheless, we assume it. There is abundant evidence to conclude that intelligence, similarly to height, being environmentally influenced, cannot be uniformly distributed; intelligence cannot present a normal bell curve.

Actually, environmental constraints on intelligence appear to affect intelligence much more severely than environmental constraints on height. We have reason to believe that an IQ test given to North Koreans and the Swiss would show a bimodal distribution as would an IQ test given to North Koreans and South Koreans because of nutritional and other environmental differences. There is similar reason to believe that giving an IQ test to people around the world would also show, at least, a skewed distribution or perhaps a bimodal distribution where there are stark differences in nutritional and other environmental factors. This is a central observation of this book; differing environmental constraints around the world yield different distributions of intelligence which suggest interventions to correspond to the environmental constraints.

The opposing and erroneous presumption, I will say myth, has plagued the study of intelligence for over a century, starting before Binet's introduction, in 1905, of IQ tests like those we know today. The scores of the earliest intelligence tests were designed to present scores that fit a normal bell curve distribution, and that requirement, based on false presumptions, carries forward to today's design of IQ tests. As will be seen, not only are IQ tests normalized, but they are renormalized every decade or so maintaining the fiction, over time, that intelligence-measures present a bell curve distribution.

That needs some clarification. Intelligence tests are not, were never, scientifically determined, but rather test-questions were and are defined, according to accepted norms of learning, such that the resulting scores of a large group of children present a normal bell curve distribution. Intelligence test scores present a bell curve distribution not because we know that intelligent behavior does, but because we assume it does.

Ceci writes,

> Because many of the early researchers in the field of intelligence were hereditarians, believing intelligence to be largely a genetically determined and transmitted trait, it was only logical [writes Ceci] that a test of intelligence should yield an approximately bell-shaped distribution. Because of this assumption, tests that did not do so were systematically refashioned, their difficulty revised, and their content reconfigured until they did yield a bell-shaped distribution. And these tests were put to one other test: They were expected to correlate highly with previous tests that also had yielded a bell-shaped distribution of scores! So, *if* the insertion of context into our notions of intelligence were to result in a non-Gaussian distribution of IQ scores, then some would take that as evidence of the weakness of such notions of intelligence rather than as a sign that the assumption itself deserved scrutiny. (p. xvi)

Ceci appropriately questions the notion that intelligence test scores should present a bell curve distribution.

Ceci also surfaces issues regarding the notion of *g*, general intelligence. He writes,

> the reader will encounter in some detail the notion of a positive manifold, that is, the assumption that performance on all intellectual tasks should be moderately intercorrelated.

If a battery of tests does not fit this pattern, some see this not as evidence that the positive manifold assumption should be challenged, but as ground for dismissal of the new battery of tests. So, in tackling the problem of the nature of intelligence, I [Ceci] soon realized that, because of the rich history of research in this area, I [Ceci] was not free to speculate at will. (p. xvi)

His not so subtle message does give one pause, does give one reason to question ingrained *truths* that may not hold up to scrutiny. There is reason, I would venture to add, always has been reason, for scientists to question previously accepted *truths* that do not comply with new evidence.

Ceci surfaces and challenges two of the key basic and deeply entrenched assumptions— myths— that have affected, and, perhaps, hindered the understanding of intelligence for over a century. A better appreciation of these notions is essential to the continuing development and comprehension of intelligence. The second issue, dealing with the notion of *g*, general intelligence, is discussed in the section regarding the *g*-factor. This is a central myth in the study of intelligence and is debunked within a few pages. The first issue, having to do with whether intelligence scores should present a bell curve distribution, is addressed immediately.

So why do we assume intelligence test scores should present a bell curve distribution? The reason appears to be based on hereditarian orientations. "Burt et al. [mentioned earlier] indicated that 'by intelligence, the psychologists understand inborn, all-around intellectual ability.'"[122] The key word is *inborn*. Intelligence was believed to be genetically determined, and accordingly presents a bell curve distribution. (Of course, even if intelligence were entirely genetically determined does not necessarily lead one to conclude that measures of intelligence should present a bell curve distribution.) The first widely accepted intelligence test was designed by Binet in 1905. There is reason to believe that the test scores presented a bell curve distribution, perhaps because, otherwise, it was assumed that the test would be viewed as invalid.

There is another reason we persist in assuming that IQ test scores and

---

[122] Hunt, pp. 340-41

intelligence-measures, in general, should present a bell curve distribution. It's *tradition*, as said so colorfully on a different subject in the Broadway show *Fiddler on the Roof*. And while I and many others view tradition with great value in some contexts, it is just as clear to many that tradition has *no* place in science. Taking this another step, why are height measures allowed to float as height changes? Probably, the reason is, because tradition never complicated the description of height; probably because the measure of height is not in dispute, is based on scientific measures, and never has been in dispute across the entire planet, whereas this is not true for intelligence.

Furthermore, there is reason to believe that if intelligence tests were to be scientifically defined they would appear as a heavily and dramatically skewed curve reflecting the fact that intelligence is largely influenced by the environment and most people on the planet live with great environmental constraints; if we could effectively measure every individual on the planet, it is not true that half of the world's population would have an IQ above 100, but rather probably, much less than 20 percent of the world's population would have an IQ above 100.

Additionally, if IQ scores could show this, as do scores for height, we would be better aware of severe human problems, for example in the Korean peninsula, as well as across the planet, and, importantly, we would have knowledge to make appropriate corrections. As now evident, the facts are unknowingly and unintentionally hidden from view by this need that IQ scores present a bell curve distribution, a process driven by tradition. Actually, this is only scratching the surface. When we talk of economic inequalities, the real issue, the underlying issue, as discussed in this book, is *intelligence inequalities*. And while there surely need not be such severe intelligence inequalities, what we have, because of environmental constraints, is intelligence inequalities far more severe than what is realized and understood today.

To be sure, intelligence equalities is not an objective, we all have different intelligence potential, but such severe intelligence inequalities, as experienced today, can and should be minimized. Intelligence inequalities partly result from the myths that intelligence-measures and IQ scores inappropriately need demonstrate a normal distribution. The dramatic environmental constraints experienced by most of the world's population and the resulting intelligence inequalities, thus, are largely hidden from view by the myths that intelligence and IQ scores are forced to appear as a normal bell curve.

We now focus on the measures of intelligence.

The myth that intelligence-measures present a bell curve distribution

A pyramid of myths, such as the one about to be described, remains a base for many false notions in society as well as a base that confuses much research in many fields. The base of this pyramid is the myth, believed by many, that several human characteristics present a normal bell curve distribution of scores. While some characteristics may present a bell curve distribution where environments are largely uniformly distributed, where the environmental factors are not uniformly distributed characteristics cannot present a bell curve distribution.

Intelligence, one of those human characteristics, has been presumed to present a bell curve distribution partly because of the myth just mentioned that human characteristics present a bell curve distribution. These issues are explained within the next few pages.

Myths spawn myths. Again, intelligence, largely determined by environmental factors, cannot present a bell curve distribution because environmental factors are not uniformly distributed. IQ tests are defined to yield scores that present a bell curve distribution because of the above-mentioned myth that intelligence-measures present a bell curve distribution. The pyramid grows. IQ tests, normalized to yield a bell curve distribution of scores, must be periodically renormalized to maintain the myth that IQ scores yield scores that form a bell curve distribution. Note, it is believed by some that this periodic renormalization is necessary because IQ scores have been steadily increasing since the 1930s, as reported by James Flynn. The pyramid needs periodic repair.

Myths of Intelligence

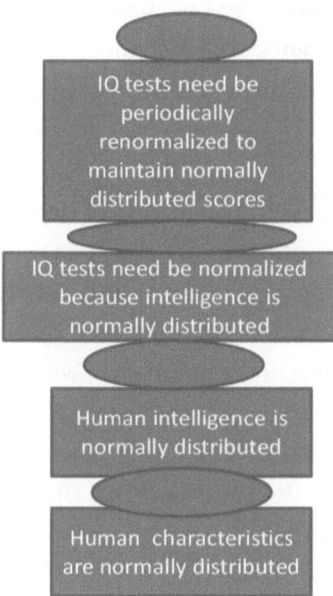

Figure 31

The pyramid in the above graphic deliberately appears to be a totem pole. This totem pole of myths has supported a wide array of false notions some of which carved in wood and then in stone by legal code resulting in great injustices. Let's look at the components of the pyramid, in turn.

Should IQ tests be defined to yield a bell curve distribution of scores?

It appears to follow that if human characteristics, in general, present a bell curve distribution, human intelligence, may likely, also present a bell curve distribution. And we persistently and stubbornly project the assumption that intelligence-measures present a bell curve distribution to IQ testing. But what if these underlying presumptions, whatever their rational foundation, is wrong? What if intelligence-measures should not present a normal distribution? What if the existence of intelligence-measures appearing as a bell curve distribution phenomenon creates a faulty base, a faulty impression? What if the renormalization of IQ test scores is arbitrarily creating and ingraining a misunderstanding of intelligence? What if that foundation supports the repeated and persistent injustices perpetrated across society, affecting not only the disadvantaged but the advantaged as well?

IQ test scores present a bell curve distribution because their designers assume intelligence-measures present a bell curve distribution. Jensen, in 1969, writes,

> test constructors make up their test of items which have rather low average intercorrelations (usually between .1 and .2) and a considerable range of difficulty levels. These two sets of conditions working together, then, yield a distribution of test scores in the population which is very close to 'normal'. So far [he writes] it appears as though we have simply made our tests in such a way as to *force* the scores to assume a normal distribution. And that [he writes] is exactly true. (p. 21)

And, I say, while true, why have we forced IQ scores to assume a normal distribution? First, intelligence, like height, does not present a normal distribution, cannot form a normal distribution, because as stated, both intelligence and height are environmentally influenced, and environmental advantages are clearly not uniformly distributed. Second, IQ tests measure both much less than and much more than intelligence. They measure much less than the entire spectrum of intelligence in that they do not measure a wide array of characteristics such as people smarts, smarts of the arts, and creativity, to mention a few shortcomings, yet IQ tests also measure much more than intelligence in that they indirectly measure nutrition, health, experiences and other environmental factors.

While IQs below 50 are probably caused by genetic influences or genetic abnormalities, IQs above 50 are largely environmental determined.

While there is no evidence indicating that intelligence-measures present a normal bell curve distribution, we do have evidence suggesting that IQ scores do *not* naturally present a normal bell curve distribution. Specifically, we know that every social class has a similar share of people whose IQ scores fall in the lowest part of the IQ spectrum, below about 50; as Jensen documented in 1969, (p. 26); it is safe to say that all groups of people include individuals with IQs below 50 who have severe mental deficiency due to pathological conditions, massive brain damage, or rare genetic and chromosomal abnormalities. Down syndrome and PKU are examples. In 1969 Jensen wrote,

> But probably the strongest evidence we have that [people with] IQs below 50 are a group apart from the mildly retarded, who represent the lower end of the normal variation, comes from comparisons of siblings of the severely retarded with siblings of the mildly retarded. In England, where this has been studied extensively, these two retardate groups are called imbecile ([people] with IQs below 50) and feebleminded ([people with] IQs 50 to 75). ... the siblings of imbeciles have a much higher average level of intelligence than the siblings of the feebleminded. The latter group, furthermore, shows a distribution of IQs that would be predicted from a genetic model intended to account for the normal variation of IQ in the population[123]. To explain the results ... one must postulate some additional factors (gene or chromosome defects, pathological conditions, etc.) that cause imbecile and idiot grades of mental deficiency. (p. 26)

Jensen conveys, also in 1969, that while an IQ below 50 probably results from pathological genetic or abnormal genetic causes, IQ between 60 and 80 are disproportionately associated with people classified as being among the disadvantaged (p. 26).

These two observations emphasize that intelligence scores should not present a bell curve distribution; that 1) there is a bubble in the IQ spectrum below a score of about 50, and that 2) the scores between about 50 or 60 and 90 are largely determined by environmental factors. This author goes further

---

[123] Jensen refers to "normal variation of IQ in the population" here, where this is merely based on an underlying assumption that he thought relevant at the time.

and indicates that scores above 50 are largely determined by environmental factors, but let's here stay with Jensen's claim.

What would explain the disproportionate representation of the disadvantaged having an IQ between about 50 and 90? Some conclude, looking at these data, that low intelligence causes one to be disadvantaged. Others conclude the opposite—being disadvantaged causes low intelligence. While each conclusion is supported by the data, additional data, as previously mentioned, provide clear evidence that improving one's nutrition, health and other environmental factors—reducing one's apparent disadvantages—increases one's intelligence, at least as measured by IQ tests.

This suggests that unlike those that have IQs below 50, intelligence derived from normal genetic material can be substantially modified with changes to the environment. Also suggested is that measurable intelligence of people across the entire spectrum of socio-economic-status should *not* presented by a normal distribution also because environmental influencing factors are not uniformly distributed.

People like to deal with data presenting a normal distribution. Considerable advantages result; predictability of change is greatly simplified, and the various standard deviations are well established. However, while there may be advantages in dealing with a normal distribution of scores, the disadvantages and injustices, to a vast number of people across societies, can be severe, and that is far more relevant.

Therefore, because measurable intelligence must reflect environmental influences, such as nutrition, health, and experiences, and because these are not uniformly distributed, it might be agreed that measures of intelligence should not be required, designed, to present a normal distribution. IQ tests, with all their limitations, should not have the additional burden of presenting scores forced to appear as a normal bell curve.

Flynn effect

Many researchers have studied the changes in IQ scores, or the likely changes in IQ scores, during the last century. Hunt reported about a half century ago that

When Cattell (1937) multiplied the number of people at each level [of intelligence] by the reproduction rate at that level and computed the new mean as an estimate of the IQ of the next generation, a procedure which assume perfect correlation between the IQ's of parents and children, he got an expected drop of a little over three points per generation, or about one point per decade. This he characterized as a 'galloping plunge towards intellectual bankruptcy', and he issued a clarion call for corrective action to preserve the intelligence of the population of the United Kingdom. (p. 337)

Clearly, none of this has proven valid.

Many studies have shown just the opposite. Hunt writes that it has been noticed that IQ scores have been increasing for nearly a century.

Perhaps the most outstanding [early] studies of this kind are the Scottish Surveys conducted by Thomson et al. (1933, 1949, 1953) of the Scottish Council for Research in Education. In 1932, nearly all 11- year-old children in Scotland got a group test of intelligence. Again in 1947, an attempt was made to reach all Scottish children who were then aged 11. The two samples numbered 87,498 and 70,805 children, respectively, and because of the national drop in birth rate, they represented 87 and 88 percent of the estimated total number of 11-year-old children living in Scotland on the two separate occasions. ... Contrary to the expected drop in intelligence over this 15-year period, the children in the 1947 sample showed a gain in mean IQ of 2.28 points, and from a statistical standpoint, this gain was highly significant. (p. 338)

James Flynn, of the University of Duneeden, for whom the Flynn effect is named, reported findings in 1978 indicating that from the mid-1930s to the late 1970s IQ scores steadily rose about three points per decade. IQ scores had thus increased during the previous five decades by about fifteen points. To place this in perspective, that is half the difference between being labeled mentally retarded and being viewed as average in intelligence. Flynn published his observations in 1984. Consequently, and subsequently, the IQ testing community has been publishing new IQ tests every ten or twelve years to maintain an average score of 100, and, to reestablish a normal bell curve distribution of IQ scores with a SD being 15 points. They normalized and then renormalized the tests several times since 1984.

Various questions come to mind. First, why have IQ scores been increasing across the last several decades? Second, why renormalize IQ tests periodically? This second question was addressed above.

Many theories have been fostered to explain the first question, why the steady and persistent increase in IQ scores since at least the early 1930s. Some argue that it is caused by improved nutrition, while some suggest these increases in IQ scores result from advances in health. Others propose that these changes are due to enhanced education. All these explanations derive from environmental improvements. As has been shown, many environmental changes cause intelligence, at least as measured by IQ tests, to increase. Others believe that these test scores have increased because of improved skills in test taking. This may have had an influence but there is relatively little evidence for this as compared to the clear evidence pointing to the influence of nutrition, health, and education.

Another possible, though highly unlikely, explanation for the increase in IQ scores is that it is caused by genetic variations—our genes influencing intelligence have *improved*. This is viewed with little credibility because while genetic changes among fruit flies and other creatures that reproduce relatively quickly, significant human genetic changes are not experienced over such a short time span. While the exact explanation for this persistent increase in IQ scores has not been definitively determined, genetic variations has been essentially ruled out.

As documented at some length earlier in this book, environmental improvements can substantially enhance intelligence in many ways; mineral and vitamin supplements, and pre-K education with nutrition are examples of ways to enhance intelligent behavior. Furthermore, according to Scarr and Weinstein, children placed for adoption at six months of age rather than twenty-four months of age demonstrate an average of 13.5 point higher IQ scores. Though this represents nearly half the difference between being considered mentally retarded and average, it does not address the first six months of life, the most significant developmental period after birth, nor does it include the prenatal developmental stage and its possible prenatal medical care. Interestingly, this study was extended by examining students not only at age seven but also again when they reached the age of seventeen. During the ten-year interim period, about half of the advantage was shown to have dissipated. Some researchers are reassured by this that intelligence is largely genetically determined, and the effects of environmental enhancements are short lived. Others conclude that intelligence can be both enhanced and diminished and that we need to better understand the reasons for both the increases and reductions in intelligence. Surely, we need to review the environmental conditions during the years between age seven and seventeen.

Of course, many environmental interventions that help people reach closer to their potential intelligence have resulted in higher intelligence among all groups in society, not only the disadvantaged, and much more could be done. More will be said about this in the final chapter.

Dozens of other ways to enhance intelligence are cited in the literature reported earlier in this book, and, importantly, these result from widely respected and often replicated research activities that are non-controversial. Michael Wines documented in the New York Times, on December 28, 2006, some of these activities and less known implications. He wrote,

> almost half of Ethiopia's children are malnourished, and most do not die. Some suffer a different fate. Robbed of vital nutrients as children, they grow up stunted and sickly, weaklings in a land that still runs on manual labor. Some become intellectually stunted adults, shorn of as many as 15 I.Q. points, unable to learn or even to concentrate, inclined to drop out of school early. ... [Furthermore, Wines writes, well] over half of sub-Saharan children under 5 lack iron, vital to developing nervous systems, the Micronutrient Initiative, a Canadian research organization, reported in 2004. They often have trouble concentrating and coordinating brain signals with movements, like holding a pencil, that are crucial to education. Another 3.5 million children

lack sufficient iodine, which can lower a child's I.Q. by 10 or more points. More than a half million suffer vitamin A deficiency, which cripples young immune systems; merely ensuring adequate vitamin A can lower child mortality by more than one-fifth. Children lacking vitamin B12, regularly measured nowhere in Africa, have impaired cognitive skills and do poorly on tests.

Thus, we know that intelligence, at least as measured by IQ tests, can be enhanced via various environmental improvements, including nutrition, health, and learning experiences. And we have reason to believe that the steady increases in IQ scores over the last several generations have been significantly caused by environmental changes.

The second question posed above, why renormalize IQ tests periodically, given the observation that IQ scores have been steadily increasing for generations? This question was partially answered in a previous section. The simple answer is easy. It's to maintain the average of IQ scores at 100, and, to maintain a normal bell curve distribution of IQ scores with which people like to deal. Neither of these reasons justifies the renormalization of IQ scores. So, why maintain the average IQ score at 100, and, why maintain a bell curve distribution for IQ scores? The answer lies in the stubborn and erroneous beliefs that intelligence measures presents a bell curve distribution, as do other human characteristics. However, the renormalization results in profound injustices placed on perhaps millions of people, for example, on the so-called mentally retarded, and limiting services given to children who would have otherwise been labeled learning-disabled.

It also reflects the federal and state laws that have been passed based on the erroneous assumption that IQ scores naturally form a bell curve distribution.

The additional question, which we will also get to later, is, what are the implications, what are the consequences, perhaps injustices, on millions of people resulting from the renormalization of IQ test scores?

There is reason to assert that measures of height and weight and many other human characteristics do not present a normal bell curve distribution, and that this is likely so because these characteristics are influenced by nutrition, health, experiences and other environmental factors that are known to be non-uniformly distributed. We do not try to normalize the curve for height after pooling the two populations of North and South Korea and finding the heights present a bimodal distribution curve. It does not even occur to the testing community to redefine the height measuring test such that its scores present a bell curve distribution and maintains a constant average. That would seem absurd. Yet, we design and periodically redesign IQ tests to maintain a constant average and a normal bell curve distribution of scores. Repeating differently, while there is evidence that measurements of many human characteristics do not present a bell curve distribution because they are influenced by environmental factors, we periodically renormalize only IQ test scores, forcing them to present a constant average and a normal bell curve distribution. Why not let there be a floating average for IQ just like for height and weight? If IQ scores, very much like height, increase, why not let them increase? And why not let the distribution of scores, and average, be what they may?

Because it is known that intelligence is both genetically and environmentally determined and we know that environmental advantages are not uniformly distributed within or among populations, we know that environmental conditions must, therefore, cause a skewing of measurable intelligence, much like it does for human height in North and South Korea. Furthermore, we have no rationale to assume or expect that just that part of intelligence influenced by genetic factors present a bell curve distribution either. We, therefore, should not require IQ test scores to present a bell curve distribution. But we do. And because we do not have any sound rationale to assume that measures of intelligence, or that part of intelligence measured by IQ tests, present a bell curve distribution, we should not be forcing renormalizing IQ tests and their scores, supposedly representing intelligence, to present a bell curve distribution, even though it may simplify statistical manipulations.

While not widely known or realized, there is considerable expense incurred by school systems and psychologists in this country as well as others to deal with this renormalization of IQ tests; hundreds, perhaps thousands, of copies of the latest test must be purchased by each testing facility (such as schools, and psychologists) to measure consistently according to the latest testing material. The substantial expense of upgrading to each successive renormalized IQ test, experienced by every school system and many psychologists, can be redirected to more appropriate expenses. Furthermore, as the IQ test scores need no longer be renormalized, we may find the necessary funds to establish a better, more complete, understanding of intelligence and therefore be able to establish intelligence tests that measure the spectrum of intelligence reflecting not only the currently limited tested factors but the factors that measure people smarts, kinesthetic intelligence, smarts of the arts, and perhaps, most optimistically, creativity.

We know that intelligence measures are differently distributed with environmental modifications. What if we let IQ scores float and thus were able to study the results? We would learn what environmental factors indirectly affect intelligence and perhaps gradually, over time, make rationally justified environmental adjustments to improve intelligence. But today, with renormalization every decade or so, the changes are not readily noticed because the testing facilities, with periodic renormalization, filter or obscure this perhaps valuable information from our awareness.

It is suggested, here, that the periodic renormalization of IQ tests be terminated. If intelligence, as measured by IQ tests, improves over the decades, then so be it. We should not raise the difficulty of the tests merely to attain an arbitrary objective of maintaining a bell curve distribution of scores or maintaining an average IQ of 100. If intelligence is improving across the population, then let its scores float upwards. Raise the average as IQ scores improve. As fewer people have IQ scores of 70 or lower, and thus fewer people are labeled mentally retarded maybe it's because *fewer* people are mentally retarded, whatever that should really mean. If more people become eligible for learning disability services, then make more funds available to support the need. If more people become eligible for the armed services (note, again, the armed services do not accept, by law, people with IQs below 80) or specific job opportunities in industry, then all to the better.

Some might claim that this gives undue advantage to the disadvantaged. Actually, it creates a more equal playing field. Some might further assert that this might inhibit society's advantages from the more worthy, whatever that might mean. I suggest that just the opposite is true. As the sea rises all floating bodies move higher. As the disadvantaged find their voice and access more of their potential, the advantaged will likely experience greater opportunity as well. The real objective is to assure a just and rational opportunity for all.

From a scientific viewpoint, one should not affect a definition based on non-scientific orientations; scientific definitions must reflect rational and scientifically defensible observations. To force IQ scores to form a normal bell curve distribution so that statistical manipulations are more easily affected lacks scientific integrity. There is no rational reason to maintain that IQ scores should present a normal bell curve distribution. Quite to the contrary, substantial evidence indicates that IQ scores should not present a bell curve distribution, should present a heavily skewed distribution.

What would happen if we did not renormalize the IQ tests every ten or twelve years? We would have fewer people labeled mentally retarded. The average IQ score would be higher, above 100. More criminals found guilty of capital crimes would be found on death row (if that remains the orientation of the powers that be in the United States).

Surely, this may disturb the IQ testing industry. Perhaps they might focus more effort on advancing the understanding of intelligence.

Several other questions come to mind but, for space, are just crisply mentioned. Do IQ scores form a bell curve just before renormalization? I think not because just after renormalization the curve is back to presenting a bell curve distribution. What changes took place to accomplish this? Was the test made more difficult at the bottom half? Does that arbitrarily keep those at the bottom, at the bottom? Does that help to fictitiously maintain the black/white IQ gap, and the disadvantaged/advantaged IQ gap? Does this observation, if true, suggest that the black/white gap is narrowing faster than the data appears to indicate? Does the periodic renormalization maintain other fictions that need to be surfaced?

We may now summarize this section by including several additional myths to the list of five myths mentioned above, namely,

6. *It is a myth that measurements of human characteristics, in general, provide a normal bell curve distribution of scores.*

7. *It is a myth that intelligence-measures form a normal bell curve distribution.*

    To the contrary, intelligence cannot form a normal bell curve distribution because it is largely determined by environmental advantages that are not uniformly distributed.

8. *It is a myth that IQ tests should be defined to yield a normal bell curve distribution of scores.*

    To the contrary, IQ tests should not be defined to yield a normal bell curve distribution of scores because intelligence does not yield a normal bell curve distribution. IQ tests should not be renormalized every decade or so, but rather should be allowed to float as does the measure of height.

Labeling people according to IQ score classifications potentially leads to great injustices

Those having an IQ of 70 or less are labeled mentally retarded, yet we know that with a little environmental enhancement one's IQ likely increases to above 70 and one no longer is labeled mentally retarded and no longer restricted from certain educational opportunities. Similar comments may be made regarding eligibility for learning disability services. This requires rethinking especially now that we know that IQs have been rising steadily for decades since the mentally retarded category was first established. While some justification existed for this category decades ago, it has significantly decreasing validity today. These considerations are further discussed in the next section.

Injustices of renormalizing IQ test scores

James Flynn's response to the question, why is there a need for IQ tests to present a normal bell curve distribution? was "The alternative [to a bell curve distribution] is to give people percentile rankings which would simply make computations more difficult." While the alternatives may present computational complexities, maintaining a normal bell curve distribution has been shown to be hurtful to many people. Every time the testing community deems it necessary to renormalize the tests, many of those who had moved up in the scale are, likely, returned closer to where they were in the scale before renormalizing. For example, someone who qualified to become a member of the armed services before renormalization, after renormalization may no longer qualify. Also, someone who qualifies for learning-disabled services before renormalization may no longer qualify after renormalization. Furthermore, all previous tests become obsolete after renormalization requiring considerable expense, by thousands of schools and psychologists, to purchase the new tests. Population measures of height are not renormalized periodically, and they are useful. Likewise measures of intelligence should be allowed to float as changes occur.

IQ tests may be an appropriate means to rank members of the population, but why do they have to be periodically renormalized. Furthermore, these test scores need not be used to identify and label anyone including the mentally retarded or the learning-disabled. Maintaining a distribution of IQ scores may do no harm if the definitions, for example, of mental retardation and access to learning disabilities services, were changed with changes in scores after renormalization. However, if the definitions do not change, as is the case, it does great potential harm, sometimes permanent and irreparable harm to many, as will be seen in the next few paragraphs.

With the steady increase in intelligence, is there any reason to suggest the labels are still valid today? If the definition of mental retardation was, perhaps, valid several decades ago, but the population has become more intelligent through whatever means, shouldn't there be a smaller percentage of people labeled mentally retarded today? But the percent of people with an IQ score of 70 or less remains exactly the same, by definition, because of the renormalization of IQ test scores.

And if IQ tests do not measure intelligence and are not adequate to identify the mentally retarded, then define a new intelligence test, or a new set of intelligence tests, along with an associated new definition of mental retardation. And while this may seem rather glib to some, it is required to end the injustices that continue to be quietly and unknowingly suffered by so many. I contend that the definitions and tests should adhere to some variation of the Hippocratic Oath briefly stated as, *"first do no harm".*

Disturbingly, the definition of learning disabilities or mental retardation does not and never has reflected a scientific understanding of these terms. Perhaps more to the point, IQ testing does not, and never has, reflected a scientific understanding of intelligence. Perhaps our understanding and definition of these terms should have been modified according to changes in intelligence tests. But they were not. Instead they remained as defined decades earlier and have consequently increasingly lost meaning, and increasingly inflicted injustices on an increasingly large number of people.

There may be reasonable questions asked about the bell curve distribution of the improved IQ scores. There is a great deal of evidence, as indicated earlier in this book, that those in the lower half of the IQ spectrum, disproportionately disadvantaged people, would be markedly and disproportionately helped by changes in nutrition, health, and experiences. Because those in the bottom half of the IQ distribution could benefit from environmental enhancement, their scores would increase disproportionately.

So, what would result if the IQ scores of those in the lower half of the spectrum disproportionately benefit from environmental improvements? The answer might be surprising. The IQ test scores of many of those who's IQs had increased before renormalization, would, after renormalization, be returned closer to where they had been before environmental improvements. Those being inhibited because of their location in the spectrum, cannot seem to escape the constraints. The renormalization acts as a subtle conspiracy to hold them back from advancing their status in society even though it is known their intelligence has steadily increased.

But, some claim that it makes little difference whether the difference in IQ distribution, after renormalization, is manifest across the spectrum, or concentrated in the lower half of the spectrum. Yet, to my knowledge, no researcher has made an argument that the increases in IQ scores, before normalization, are disproportionately noticed among those in the upper half of the spectrum.

Surely, the best way to provide evidence, here, is to get the data that was used by James Flynn or that continues to be used by the IQ testing community. Via personal communication, I learned that James Flynn no longer has the information, and to the best of my knowledge, the IQ testing community has not made this information public. I suggest that that data be made available not only in justifying the renormalization of past IQ test scores, but more clearly to justify recent and continuing renormalization of IQ test scores. Surely, the testing community will have adequate opportunity to disprove these observations with data. And though analytic rationale just might be adequate to make the necessary observations, the stated rationale may be wrong; however, it seems too rational to be refuted without the real data.

Interestingly, some people have argued that the disadvantaged/advantaged IQ gap, in general, and the black/white gap in IQ scores, in particular, has hardly changed in the last several decades. They suggest that if intelligence is influenced by the environment, and as the environment of the disadvantaged has improved for the last several decades, why has their relative position in the IQ spectrum hardly changed? Perhaps, they assume, the reason is subtly concealed in the realization that intelligence is largely genetically determined. On the other hand, perhaps the answer is partially found in the renormalization of IQ test scores every decade or so. As intelligence rises, the tests are made more difficult, perhaps especially for those in the lower half of the spectrum, and intelligence scores are returned closer to their previous place in the spectrum according to the now more difficult test. Perhaps the system is unknowingly rigged against the improvement of the disadvantaged in society. Perhaps there is an unintended consequence that is so subtle that not only is the general public unaware but even the experts developing the new tests are also unaware.

But, one might validly insist, if the environment has gradually changed during the last several decades, and that has disproportionately benefited the disadvantaged, then why have they not moved up in the scale? Truth be told, as shown in Chapter 5, the average IQ of the black population in the United States has increased by about five points more than the white population; the IQ gap has collapsed by about 1/3. Furthermore, there is reason to believe that the gap may have shrunk more had renormalization of IQ scores been terminated.

To summarize, what are the injustices caused by renormalizing the IQ scores? Examples include limited access to the armed services. But not only is entry into the armed services made increasingly difficult with each renormalization of IQ test scores, but this is also true for industry in general, even though industry is not allowed, by law, to restrict a job from anyone because of the results of an IQ test. Additionally, constrains are placed on people with learning disabilities to receive extra educational services—they must have higher intelligence than was required at the time the relevant laws were passed regarding learning disabilities. Also, many are being labeled mentally retarded with higher intelligence, though similar IQ scores, than when the definition of mental retardation was first determined. Decades earlier they would have been viewed as near average. Also, an individual with an IQ score of, say, 70 may be relegated to special classes for students viewed to be incapable to achieve in regular classes. That student is restricted from educational opportunities that would have been available had the test been taken several years earlier. As IQ test scores are repeatedly renormalized, they are, by definition, made more difficult, making it increasingly more difficult for many students to receive the education to which they are entitled. This may also partially explain why the number of children identified with learning disabilities has been steadily increasing over the years, though it surely does not completely explain that concern.

Furthermore, regardless of one's view of capital punishment, the law states that people who are mentally retarded—people with an IQ of 70 or lower—cannot be given the death penalty. Rather, their punishment may be life in prison. Those that argue against the death penalty are rather pleased with this, but those that favor capital punishment seek the re-sentencing of those who took the IQ test just before renormalization. Periodically, cases are heard by the courts because of just this issue.

One such case occurred some time ago. James Flynn and others were called as expert witnesses in cases involving such capital crimes. Flynn wrote an article on the subject in 2006. The complication in these cases results from various issues, one of which is when, the date, the IQ test was given that determined the prisoner to be viewed as mentally retarded. If an IQ test had been given just before renormalization, then the prisoner might be viewed as not mentally retarded, whereas, if the prisoner had been tested just after renormalization, he or she might more likely have been determined to be mentally retarded.

So, what are the reasons for the periodic renormalization of IQ tests? Several reasons come to mind, essentially all of which are less than adequate. They are based on the general understanding, or misunderstanding, of intelligence—an ingrained view that intelligence is relatively fixed, a myth, and that there is not much one can do to change it, another myth. The facts, as shown earlier in this book, show that intelligence constantly changes—changes substantially with nutrition, health, experiences, and other environmental factors. There is much, very much, that can be done to substantially increase intelligence, at least as measured by IQ tests.

Possibly, the most significant injustices stem from the unfounded view that intelligence cannot be enhanced. This view discourages the implementation of interventions resulting in millions of people in the United States alone, billions on the planet, to remain unnecessarily in poverty.

IQ score regression-toward-the-mean

The notion of regression-toward-the-mean regarding IQ scores conveys that the IQ of the offspring of a couple will tend to be between the IQ mean of the population and the IQ average of the parents; if each of the parents of an offspring has an IQ of, say, 125 and 115 respectively, for an average of 120, the offspring would likely have an IQ of somewhat less than 120, somewhat closer to the population mean of 100. Jensen argues that just like height, the IQ of offspring regress predictably to the population mean. Is this claim valid? What would justify it?

Some basic observations need to be considered. Human characteristics such as height and intelligence, as said, are influenced by both genetic *and* environmental factors. (Recall the height of the North and South Koreans are significantly influenced by nutrition and other environmental factors.) While some may claim that genetic factors influence regression-toward-the-mean, few would claim that environmental factors determine regression-toward-the-mean.

Let's first discuss the genetic influence. If genetic factors affect regression-toward-the-mean, then after many hundreds of generations, the height, or intelligence, of us all, by now, would have arrived at the population mean. Obviously, that has not happened. Therefore, one must question whether genetic factors cause this regression-toward-the-mean. More likely, humans have a similar, though not identical, genetic cognitive capability. Incidentally, one may claim the same for hamsters, albeit they appear to support a different spectrum of genetic cognitive capability. There is no reason, no genetic evidence, to suggest that human or hamster genetic cognitive capability has or will regress toward their respective means.

As to the environmental effect on regression-toward-the-mean, we know that improving or reducing nutrition, health, and other environmental conditions enhances or limits height as well as intelligence, but only in one direction; improvements increase intelligence whether the starting point is below or above average, and reductions in environmental conditions reduce intelligence whether the starting point is above or below average. Certainly, some people experience improved environmental conditions while others experience reduced environmental conditions, but they do not balance each other because there are many more disadvantaged in this world than advantaged.

Also, Jensen's claim for regression-toward-the-mean of intelligence, or more precisely of IQ scores, assumes a strong influence of genetic influence, whereas, especially among the disadvantaged – most of the world's population – the environmental factors influence intelligence far more significantly than do genetic factors. Thus, environmental factors, if improved, would, likely, outbalance, far more than outweigh the possible genetic effect of some tendency of regression-toward-the-mean.

Furthermore, IQ scores reflect both much less than and much more than the array of intelligence; IQ tests measure much less than the array of intelligence because they do not measure creativity, people smarts, talents, intelligence in the arts, while IQ tests measure much more than intelligence because they measure environmental influences. Therefore, a conclusion of regression-toward-the-mean concerning intelligence cannot be rationally derived from just IQ scores. Few conclusions concerning cognitive genetic factors influencing intelligence can be rationally derived from just IQ scores.

While some studies have shown that human height tends to regress-toward-the-mean, no data has ever been shown, to my knowledge, that intelligence scores regress-toward-the-mean. It has merely been assumed.

Thus, without data showing the opposite, it may be safely maintained that

9. *It is a myth that IQ scores or intelligence regress-toward-the-mean.*

Is *g* real?

The notion of **g**, which, currently, represents *general intelligence*, implies the existence of one, or, more likely, many genetic factors that determine intelligence. The very existence of **g** was recognized and introduced over a century ago by Charles E. Spearman. The notion was originated during a period in which many viewed intelligence as determined at conception, determined by genetic factors. Recall Burt's definition of intelligence being "inborn, all-around, intellectual ability". Over the century, the concept of *g* has evolved and matured. The discovery of *g* has been viewed, by some, but *not by this author*, as possibly the most significant event in the history of psychology and, more specifically, the development of intelligence theory. Jensen and many other researchers have devoted a great deal of effort to understanding the *g*-factor. Some have focused on finding the genetic factor(s) that determine *g* with limited success. I contend this is partly because these studies are based on IQ studies that are, by definition, confounded by environmental considerations, but likely more importantly because many researchers continue to believe that intelligence is mostly determined by genetic factors. Truth be told, there is reason to believe that *g* is determined mostly by various environmental factors – is hardly influenced by genetic factors.

The notion of *g* was recognized upon closely examining intelligence test scores—the correlation of various IQ test scores and the correlation of sub-test scores. Interestingly, if the tests had been made of other human activities, such as sports, the arts, as well as intelligence, one would have discovered that the notion of *g* applies not just to intelligence but to all human activities. Furthermore, as has been shown and is widely acknowledged, scores of IQ tests, behavioral tests, are at least significantly, or, largely influenced by environmental factors. Therefore, the *g*-factor largely results from environmental influences. The evidence suggests that *g* should not represent *general intelligence*, but, more appropriately, it should represent general nutrition, general health, or, more generally, *general well-being*—that the few most influential underlying factors of *g* may relate to nutrition and several other environmental factors. Though the notion of *g* is real, it more likely refers to *general well-being or general capability* rather than *general intelligence*. This may explain why a person that is effective in one area of human endeavor is probably also effective in others. It's not because of innate intelligence but because of well-being.

With this as background, the notion of g has a substantially different relevance, a much wider application, than previously thought; because various aspects of only intelligence, as measured by IQ tests, were examined, it was assumed that g represented the underlying commonalities across differing aspects of intelligence. But as the underlying, more basic, commonality is shown to be well-being (with perhaps yet-to-be-found some genetic commonality), the notion of g is broadened; more generally, g introduces the notion that if an individual is capable in one human activity he or she is likely capable in several activities. The notion of *g* does not merely reflect a correlation between different aspects of intelligence, as previously assumed, but far more significantly, *g* reflects a correlation between many differing human activities; certainly, the evidence supports the high correlation between one area of intelligence and another area of intelligence, but more importantly, it also reflects a correlation between intelligence and skiing, between intelligence and playing tennis, and between playing tennis and skiing. The correlation in all four cases is claimed, here, to be caused by well-being, the underlying common strength. It may include some reflection of intelligent behavior, however, more significantly, it reflects well-being. *Thus, seeking cognitive genetic factors that support the g-factor has proven rather futile because the g-factor does not reflect only intelligence, but more generally, it reflects well-being.*

While there is no evidence that g results from cognitive genetic factors, there is abundant evidence that g results from environmental factors such as nutrition and health that affect well-being. Incidentally, there may be evidence that some genetic factors support g, but not merely cognitive genetic factors.

What are the implications of *g* when concluded to be largely environmentally determined? As will be seen, the ramifications on sociological considerations and the effects on research are striking. They redirect the various dependent approaches in both education and research. Thus, in contrast with what Ceci wrote, shown earlier in this chapter, in tackling the problem of the nature of *g*, "I [Ceci] soon realized that, because of the rich history of research in this area, I [Ceci] was not free to speculate at will". In contrast, this researcher suggests that because intelligence is significantly, I say largely, determined by environmental factors, that the *g*-factor is likely caused by environmental factors. *Thus, g should stand for general well-being or general capability – not general intelligence. This places an entirely different understanding of the notion of g and its implications.*

It is concluded that

10. *It is a myth that the g-factor is one or more cognitive genetic factors. While g is real, it is much more general than just applying to various aspects of intelligence. g should stand for general well-being, and thus, has a much wider relevance than currently thought. g is determined by*

well-being, general well-being, which is determined by nutrition, health,
and other environmental factors; the more well-being, the more one's
*general capability. The notion of **g** has much broader and much more*
obvious implications including physical capabilities.

IQ testing considerations

Researchers and educators hold a spectrum of views about IQ testing. For example, Loehlin writes,

> Although psychological tests have not been without their detractors, even at the outset, there is little doubt that such tests represent one of the most significant technological accomplishments of the social sciences. (p. 5)

There is reason to question this judgment. John Garcia of the Department of Psychology at UCLA writes,

> Richard J. Herrnstein (1971) proclaims that the measurement of intelligence is psychology's 'most telling accomplishment to date' p. 45. [Apparently, Herrnstein is in agreement with Loehlin. Garcia then adds,] What a stunning indictment [of psychology]![124]

This author agrees with Garcia; IQ testing used within its initial intent has value, but when generalized and used to provide evidence regarding a wide range of psychological considerations, has hampered and hindered progress toward understanding intelligence, and resulted in great injustices.

The first IQ tests, designed by Binet before 1905, were based on teacher feedback directly supporting the limited objective to identify which children were best prepared for education and which needed assistance—hardly based on proven theory of intelligence.

---

[124] Eckberg, p. ix

Eysenck (1962) states that 'intelligence tests are not based on any very sound scientific principles' (p. 8). Others have admitted that 'none of the existing scales are based on any recognized theory of intellectual development' (Warburton et al. 1972); and David Wechsler (1971) admits that even where theories are quite different, the general types of items used are limited, because statistical, rather than theoretical, requirements are considered of paramount importance. (p. 23)

Surely, these tests identified children needing help to become educated, as was their initial intent, but they did not advance a basic understanding of intelligence—undoubtedly an accomplishment but hardly the accomplishment implied. Again, there is reason to claim that IQ testing has severely delayed progress toward understanding intelligence, a core notion that underpins psychology and much of the social sciences. In this regard, one can make the case that IQ testing has proven to be among the least significant developments that happened to psychology, and that its misuse has significantly hampered progress in many disciplines for about a century.

IQ tests have real but only limited value. While they measure school ability, according to its original intent, IQ tests measure only some aspects of intelligence; they hardly contribute to an understanding of intelligence during the perinatal stage, and IQ tests only somewhat measure the growth of some aspects of cognitive ability, and only measure the growth of some aspects of knowledge such as vocabulary. While knowledge is not intelligence, it certainly contributes to intelligence, to intelligent behavior.

Though IQ tests measure a subset of intelligent behavior, essentially all behavior reflects not only genetic factors but environmental factors such as nutrition, health, and learning experiences. And these influences, currently, cannot be distinguished at the time of testing.

IQ tests appear stable because they are designed to be age-independent. But intelligence, for example that of Einstein, reflects the collective influences of one's information processing capability and one's accumulated information, in addition to emotion, nutrition, health, motivation, orientation, and creativity. The stability of IQ scores assumes the consistency of its influences according to the expected age-dependent genetic and environmental influences at each age.

Are IQ tests culture biased?

Many have argued that IQ tests have built-in culture biases. The test designers are aware of this criticism and have attempted to correct that problem over the years. Having said that, I feel that it is difficult to eliminate all cultural biases when tests include language, reflect dialect, and general knowledge gained from experiences. So long as tests measure some aspect of learned behavior, they will be culture biased. However, this criticism has less validity today than it had in the past.

A different orientation to intelligence testing is needed.

Several researchers have recognized the need to develop a new form of intelligence measurement. Sternberg, then at Yale University, introduced a theory underpinning a different orientation to intelligence testing. Additionally, some intelligence measurement tests have been developed such as the Raven's Matrices. None of these tests have received consensus support. IQ testing using, for example, the Wechsler Intelligence Test, continues to be viewed by many as the most operationally effective IQ test. This area needs much more focus.

More on the question whether IQ scores are stable.

We know intelligence cannot be stable because, first, intelligence is influenced by environmental factors, which are not stable and, second, intelligence grows with experiences and is age-dependent.

IQ scores, on the other hand, Hunt observed, are designed to be stable.

> Binet and Simon (1905) noted … improvement of IQ with age immediately, and it led them, in their 1908 revision, to grade their tests according to age, and, in their 1911 revision, to introduce the concept of mental age. A test was considered to be typical of a given age if approximately

261

three-fourths of the children of that age passed it successfully. Wilhelm Stern (1912) should probably be given the
credit for suggesting the intelligence quotient. (p. 15)

This resulted in IQ test scores appearing stable.

So, while IQ scores are stable by design, it should not suggest, as it has
for many, that intelligence is stable. For IQ, which is age-independent, to
remain stable, intelligence, which is age-dependent, must increase according
to age. Stated differently, IQ is stable because both the environmental factors of a child as he or she matures are often relatively unchanging, and, the
expected growth of knowledge, for large groups of individuals, is also rather
predictable. However, this does not necessarily hold for the disadvantaged,
who occasionally experience an increase or decrease in status, and thus
experience enhanced or diminished environmental conditions, and thus
more or less effective formal and informal learning, and thus decreasing or
increasing IQ during different periods over the years. Also, the IQ of the
disadvantaged, who experience constraints on nutrition and other environmental factors, decreases steadily through the developmental stages. So,
again, while IQ scores are age-independent by design and therefore appear
relatively stable for the advantaged, they are not stable for the disadvantaged.
Intelligence is age-dependent and therefore is expected to increase with age.

Is IQ testing a valid measure of intelligence in adulthood?

Although IQ tests remain stable for groups of people, and while IQ tests are
age-independent, through about age twenty, they become less effective to
measure intelligence as some individuals move into and through adulthood.
Furthermore, intelligence grows not only through childhood but through
adulthood as well. Adults often become better able to deal with human interrelationships as they age; that is not measured by IQ tests. They become
better informed on various subjects, but these, often, are also not measured
by IQ tests. A 72-year-old may well behave more intelligently than a 22-year-
old in several areas. So, while IQ tests do not measure these growing abilities, intelligence does advance.

Historically, intelligence has been viewed by some as being determined at conception and does not grow. But as is well known now, and

demonstrated in this book, intelligence grows with every formal and informal learning experience which continues throughout most of one's life. Thus, it is concluded that

11. *It is a myth that intelligence does not grow throughout most of one's life. It grows well into the last stages of life.*

Is intelligence potential fixed or open-ended

While there is reason to believe that we are conceived with a potential intelligence, a potential that appears essentially fixed, intelligence is actually open-ended. The more knowledge, the further the potential intelligent behavior moves out; with the introduction of immediate access to a world of electronic information, intelligent behavior increases and will likely continue to increase. Therefore, it is concluded that

12. *It is a myth that an individual's potential intelligence is limited at conception. Potential intelligence is open-ended and grows with every information gained from formal and informal learning experiences including from the, now, apparently open-ended electronic information bank.*

Summary

In net, everything indicated in this chapter effects the need to understand intelligence, its natural growth, and its enhancement. I debated whether to include this chapter and finally decided that it is pertinent to this book because it relates to convincing people that intelligence can be substantially enhanced, increasingly and substantially depending on one's place in the ranks of the disadvantaged and advantaged. Myths need to be revealed, debunked, and agreement must be reached as to what notions should finally be moved into the field's mythology. This is necessary so that understanding intelligence may proceed without hindrance.

Following is a list of myths that this author suggests need to be moved into the field's mythology;

1. *It is a myth that intelligence among disadvantaged and advantaged humans is largely determined by genetic factors.*
2. *It is a myth that differences in intelligence between groups of individuals are even partly genetically determined.*
3. *It is a myth that the disadvantaged are genetically more limited in potential intelligent behavior than the advantaged*
4. *It is a myth that blacks are genetically more limited in potential intelligent behavior than whites*
5. *It is a myth that intelligence is stable over time.*
6. *It is a myth that measurements of human characteristics, in general, present a normal bell curve distribution of scores.*
7. *It is a myth that intelligence-measures should present a normal bell curve distribution.*
8. *It is a myth that IQ tests should be defined to yield a normal bell curve distribution of scores.*
9. *It is a myth that IQ scores or intelligence measures should regress-toward-the-mean*
10. *It is a myth that the **g**-factor is one or more cognitive genetic factors. The **g**-factor is more likely environmental factors and, to a lesser degree, perhaps, but not necessarily, genetic factor(s), as well. **g** should stand for general well-being or general capability – not general intelligence.*
11. *It is a myth that intelligence does not grow throughout most of one's life; actually, it grows well into the last stages of life.*
12. *It is a myth that an individual's potential intelligence is limited at conception. Potential intelligence is open-ended and grows with every information gained from formal and informal learning experiences including from the, now, apparently open-ended electronic information bank.*

Research considerations and implications

Realizing that intelligence is substantially determined by the environment, and, what's more, differences in intelligence among most humans are largely determined by the environment, suggests the need to reexamine and question fundamental notions of intelligence and various research topics with which academia has been struggling for over a century. This chapter suggests the need for a new paradigm underlying the understanding of intelligence—a paradigm unencumbered by the persistently stubborn myths debunked above; a paradigm freely permitting and encouraging the introduction of notions that build a firm foundation upon which to construct a rational and sound understanding of intelligence, including its measurement, its growth, and its influencers, and thus to more effectively support the many fields of essentially every human endeavor.

Intelligence must be demystified.

As Jensen and others have said, there are as many definitions of intelligence as there are experts defining it. Some emphasize the existence of a single or several genetically derived entities—the *g*-factor—that they claim determines intelligence. Others, for example, Howard Gardner of Harvard University, claim there are many types of intelligence—multiple intelligences—that have different genetic derivations. Some assert, with sincerity, that intelligence is whatever IQ tests measure. One of the few things on which psychologists agree is that IQ tests do not measure intelligence so much as they measure current educational potential. In fact, IQ tests do not measure genetically determined cognitive structures. They can only measure behavior, albeit intelligent behavior, but only, then, a sliver of intelligent behavior. IQ tests do not measure creativity, or most of the intelligences Howard Gardener described, just to mention a few of its limitations.

We know intelligent behavior is based on cognitive structures, the well-being of those constructs, as well as accumulated information. Intelligent behavior is also significantly affected by various emotional factors such as stress. Cognitive structures function better when experiencing well-being. We also know that accumulated information is entirely environmentally determined—our accumulated information starts out as a blank slate. Surely, there are some instinctive behaviors that appear genetically determined, but among humans, accumulated information appears far more significant in determining what we view as intelligent behavior, than instinctive behavior. And while we know that IQ tests do not measure cognitive structures directly, we know that IQ tests measure behavior that surely is partially determined by genetically driven cognitive structures as well as environmentally determined accumulated information, both significantly influenced by neuronal well-being which, in turn, is influenced by environmental factors such as nutrition and health.

Nonetheless, while we have no generally agreed definition of intelligence, we have assumed several ingrained notions of intelligence. None of us reach our potential intelligence partly because of the environmental constraints we all encounter. We are unhealthy at least part of our lives; we are perhaps undernourished for at least short durations, or we are not exposed to experiences that support our genetic orientations—if Mozart had not had access to a clavier, we could not now enjoy the fruits of his creativity.

We all reach some level of intelligence below our potential, most of us well below our potential. The more advantaged, the more closely one may reach one's potential. And as our environment constantly changes, from moment to moment, from year to year, we achieve, accordingly, more or less of our potential intelligence.

Several basic questions need to be resolved in a more directed way, much like the processes used in the Genome Project. The topics might include disambiguation of the currently unknowable mix of genetic and environmental influences on intelligence, identifying the genes and their relationships that influence intelligence, launch an organization with many *coordinated* activities, located all over the world, to establish an understanding of underlying intelligence structures, define a measurement tool to measure those genetic factors that affect intelligence at conception, and, understanding creativity. Those that have devoted much time and effort to understand intelligence know that this list represents only a partial catalog of basic questions. Many important topics are not, but should be, included, some of which may be viewed as more basic and important. We need to demystify intelligence.

I once asked a group of young men, "What is the most important part of your body. Some answered spontaneously, "It's what's between my legs." I responded, "Wrong! It's what's between your ears." Arguably, the most urgent human activity is to "know thyself". A significant part of that is to understand the mind, to understand intelligence. Some time ago, President Obama initiated an effort to better understand intelligence. This needs to be expanded with funds comparable to the funds provided for the Genome Project. The questions are many. The confusions are profound. The potential value is immense.

Perhaps a new conclave should be convened such as the one chaired by Ulric Neisser in 1998 and the one held a half century earlier than that, to discuss an appropriate way to develop a scientifically supportable intelligence paradigm. The above-mentioned myths should be discussed and, with agreement, placed into the archives of the field along with phrenology. A foundation should be established upon which new research may be identified, coordinated, and more productively developed. This would not only advance and satisfy intellectual curiosity and advance the understanding of arguably the most central human feature, intelligence, but would help us understand social issues and human interrelationships, firmly build societal structures, and minimize social injustices.

Stephen Hawking[125] wrote,

> In order to talk about the nature of the universe and to discuss questions ... [perhaps regarding intelligence], you have to be clear what a scientific theory is. [Hawking wrote,] I shall take the simpleminded view that a theory is just a model of the universe, or a restricted part of it, and a set of rules that relate quantities in the model to observations that we make. It exists only in our minds and does not have any other reality (whatever that might mean). A theory is a good theory if it satisfies two requirements. It must accurately describe a large class of observations on the basis of a model that contains only a few arbitrary elements, and it must make definite predictions about the results of future observations. (p. 10)

Here we present a model without bias, describing a large class of observations, and making definite and demonstrable predictions about the results of future observations.

---

[125] Stephen Hawking. (1998). *A Brief History of Time; Updated and Expanded Tenth Anniversary Edition.*

CHAPTER 8

Summarizing evidence provided in various chapters of the book

In this chapter we review and summarize just the evidence presented in previous chapters. We, a) recap nutritional interventions that reduce environmental constraints that, accordingly, raise intelligence, b) crisply review how intervention costs are essentially immediately, in parallel, offset by savings in related areas, c) briefly summarize the conclusion that, as we raise intelligence we substantially reduce many societal problems including school dropout rate, under-employability, criminality, welfare, and poverty, d) examine interventions to reduce poverty versus relief services to reduce the pain of poverty, e) convey current models showing reasonableness of interventions, and f) crisply reemphasize myths regarding intelligence that need to be placed into psychology's mythology.

We first focus on nutritional interventions, the most effective, the most readily applied, probably the least costly, that increase intelligence of the disadvantaged, and also strongly effect health.

Recapping interventions that increase intelligence by reducing environmental constraints

For most of the last century, an IQ gap existed between the advantaged and disadvantaged. Interestingly, while studies largely focused on the IQ gap between blacks and whites, the solutions apply similarly to the IQ gap between the advantaged and disadvantaged. That gap has been shown to be 15 points, but since the 1990s, evidence shows that the gap has decreased to about 10 points[126]. Importantly, these reported IQ gaps are averages, and thus, are at least somewhat misleading; the IQ gap is 4.5 points at age four and gradually increases to 15.5 points at age eighteen, resulting in an average of 10 points, misleading claimed to be the gap at all ages. Examining the evidence shown in previous chapters, while using the reported IQ gaps at age four through eighteen, permits a much greater understanding of the IQ gap between the disadvantaged and advantaged, regardless of race.

Because there are limited tools to determine intelligence before the age of four, we first recap the evidence explaining the IQ gap at age four. We then discuss the gradual and steady increase in the IQ gap from the age of four to the end of high school. The explanations of the initial IQ gap at age four include,

- The disadvantaged have twice as many premature and LBW babies as do the advantaged, regardless of race. LBW *related deficits contribute 1.3 points to the IQ gap at age four; that, alone, accounts for more than 25 percent of the IQ gap currently reached by age four.* See Chapter 5 for evidence.

- Providing disadvantaged pregnant women, regardless of race, with mineral and vitamin supplements increases their offspring's IQ by an average of 8 points; this, alone, decreases *the IQ gap between the advantaged and disadvantaged by another 1.3 points and* explains another 25 percent *of the IQ gap currently reached by age four.* This intervention would also reduce the frequency of LBW babies and thus additionally reduce the IQ gap at age four. Notably, doctors prescribe mineral and vitamin supplements to pregnant women during prenatal care. Significantly, the disadvantaged get prenatal care less frequently than do the advantaged. See Chapter 5 for evidence.

- Providing mother's milk to babies increases their IQ by an average of 7 points. The disadvantaged, regardless of race, feed their babies mother's milk considerably less than do the advantaged. This accounts for another about 22 percent of the IQ gap between the disadvantaged and advantaged at age four. See Chapter 5 for details.

---

[126] Dickens, W. T. & Flynn, J. R. (2006a); Nisbett, R. E. (2005).

Just these three interventions explain most, nearly 75 percent, of the IQ gap at age four between the advantaged and disadvantaged – black or white. Additionally, it is reasonable to observe that these nutritional constraints limit the effectiveness of formal and informal learning experiences and thus explain the reduced development *and* growth of intelligence by age four. So, essentially all the IQ gap at age four between the advantaged and disadvantaged, regardless of race, is explained by environmental factors.

We now turn to nutritional considerations causing changes in the IQ gap between the advantaged and disadvantaged that develop between age four and through high school and even college. The IQ of the advantaged remains essentially constant through childhood, with a strong correlation (0.70). In contrast, the IQ of the disadvantaged, including blacks, decreases gradually and steadily through childhood by about 0.6 IQ points each year from age four through age twenty-four amounting to 12 points (20 x 0.6). What would explain this steady decrease in IQ? Of course, this steady decrease in IQ among the disadvantaged along with the essentially unchanging IQ among the advantaged results in the correspondingly steady increase in the IQ gap. While there is no evidence to explain this genetically, environmental issues such as limited nutrition, poor health, and accordingly less effective formal and informal learning experiences, largely explain this. Several researchers[127] have reported that providing mineral and vitamin supplements, through high school increases intelligence, as measured by IQ tests. Thus, we know that enhanced nutrition increases intelligence in the moment which then permits more effective learning experiences, thus, increasing growth of intelligence. Both observations prevent the decrease of IQ, and accordingly prevent the IQ gap from developing.

This needs some qualification; while there is no evidence, none whatsoever, to show that the IQ gap between blacks and whites, or the disadvantaged and advantaged, is genetically determined, there is no evidence to show that the IQ gap is not genetically determined either. Some psychologists may take comfort with this observation. Having said that, however, there is overwhelming evidence that the IQ gap between blacks and whites, and between the disadvantaged and advantaged, is environmentally

---

[127] For example, Herrnstein and Murray.

determined; many studies, all juried, some replicated, collectively show that environmental factors cause the entire gap. Chapters 4 and 5 provide details.

One may wonder why there is a continuing need to propose genetic hypothesis to explain this IQ gap. The study of intelligence must scientifically focus on reality based, evidence based, and reason-based approaches. The study of intelligence should not be distracted by less than proven, perhaps preconceived beliefs.

I am reminded of a similar but subset of interventions in a somewhat different context, as may be recalled from a story told earlier in this book. While traveling in Tanzania, my wife and I had the opportunity to visit several rural schools. We learned that more than 90 percent of the children had not graduated to secondary education until free breakfast and lunch was introduced. Now nearly all the children in those schools move on to secondary schools. Without adequate well-being, in the form of nutrition, education has limited value.

Though there were about sixty-five children in the classrooms with three to a desk, little tables, the children seemed happy and driven. While there, my wife and I had the opportunity to circulate in the classroom and chat with the children; several children repeatedly encircled me. Speaking some English, they asked me my name. I told them, with a twinkle in my eye, "my name is Barack Obama". They smiled and then belly laughed. School may be a drag for many children, but the central point here is, when they are well nourished the chances are much greater that they will be happy, more focused, and more successful in school. Children are naturally curious and driven to learn both formally in school and informally in other settings.

Although, this book hardly focuses on health implications effecting intelligence, there has been enough discussion to rationally see that, like limited nutrition, poor health including obesity, stress and the related medical issues, effect intelligence in the moment and its steady growth over time.

As to education, we have indicated several interventions including a) providing earlier quality formal and informal learning experiences such as early schooling for three-year-old children. Also, b) extend formal and informal learning experience throughout the year, encouraging year-round education, and school break educational opportunities for all children. Additionally, we encourage c) parenting education in high school and beyond for all, especially for prenatal and perinatal care but also for parenting young children and young adults. This is intended to help potential parents through this challenging time and preparing parents for their new life after birthing. We spend a great part of our life parenting, but usually with no more knowledge than our parents instilled in us. And while we know our parents made mistakes, unreasonably we seek little by way of improving our approaches to parent, arguably, for many of us, the most important activity of our lives.

Taking this another step, we know that parenting has a significant effect on the well-being and intelligence of offspring. Parenting training provides the knowledge not only to assure offspring adequate nutrition and good health but also encourages talking to children, and, perhaps more importantly, listening to children, and helping rather than doing for children. Parenting training also provides the underpinnings for self-confidence, motivation to achieve, and enthusiastic stimulation for learning. While we devote much time and effort in school to study languages, science, and math, and sometimes the arts, parenting education rarely is included in the school curriculum. Parenting is apparently to be absorbed from home experiences which, as we know all-too-well, may be less than optimum or effective.

There are strong reasons to change this. Parenting training must be recognized as vital to society, to both the advantaged and disadvantaged. Parents learn to parent on the job and then after developing some experience, move on to other activities. Little value is passed on, or studied, or conveyed appropriately to the next generation.

Incidentally, parenting training in high schools is available in some but very few locations, as mentioned earlier in this book. Parenting training programs exist in Wake County, North Carolina and various other parts of the country (see a program called *Education for Successful Parenting,* run by Randi Rubenstein).

While implied throughout this book but not explicitly fostered, tutoring and mentoring programs in schools should be dramatically expanded. Many, of all ages, and an increasing number of retired folk, would be overjoyed to help children; the lives of both children and retirees would gain additional meaning.

It is the collection of these interventions – to enhance intelligence and thus decrease poverty – that is the central motivating force to have written this book.

Note that there is no suggestion here to change educational processes, though there is clear evidence that that might well be fruitful; there is evidence, not discussed in this book, to show that it would be effective.

Intervention costs are entirely offset by immediate savings in related areas

Some people will claim that these interventions are costly. Actually, the costs of the suggested interventions are entirely and *immediately* offset by savings in closely related areas. In fact, the costs are trivial compared to only the short-term benefits to society, not to mention the long-term benefits.

Some of these related areas are national health care costs which are disproportionately affected by those in poverty. This results not only from poor nutrition but also from poor health care and delayed health treatment. Though these observations are more severe in the United States than in most other developed countries, they are far more severe in the underdeveloped countries. However, as poverty decreases so do much of health care costs.

Though inadequate nutrition typical among the poor not only constrains intelligence, it causes a wide array of serious medical conditions such as diabetes, stress related medical conditions, high blood pressure, and much more, not only effecting immediate health but producing chronic medical conditions. For example, the disadvantaged have twice as many LBW babies as do the advantaged. Moreover, LBW babies born to the poor are, on average, lighter than the LBW babies typically born to the advantaged; the initial hospital charges for lighter LBW babies are significantly costlier than for the heavier LBW babies born, on average, to the advantaged. Moreover, the long-term effects, typical of very low birth weight (VLBW) babies are far more detrimental not only on intelligence but on many medical conditions. These are hardly measurable. The evidence is more fully described in Chapter 4.

Providing mineral and vitamin supplements to disadvantaged pregnant women not only raises the IQ of their offspring but, more relevant to this section, decreases the incidence of LBW babies, saving enough money in LBW medical services to support providing mineral and vitamin supplements to all disadvantaged children throughout their school years. This notion was first introduced in Chapter 1 and expanded in Chapters 4 and 5. In net, nutritional interventions provide both long-term and short-term solutions with immediate savings supporting the entire cost. These interventions alone would largely eliminate the IQ gap between the advantaged and disadvantaged regardless of race.

And there is much more. Adequate nutrition not only improves intelligence but also improves general health. The savings in health care costs, in addition to the savings in medical costs for LBW babies discussed just above, probably would pay much of early schooling starting at age three, likely the costliest intervention. We have focused on only one *health-care* cost – the initial medical care cost of LBW babies, which substantially results from the effects of poverty. Additionally, we could probe the cost of diabetes, obesity, high blood pressure, stress related medical problems, and many more health issues[128] substantially more prevalent among the disadvantaged than the advantaged. Furthermore, evidence shows that the poor are much more frequently ill and much more severely ill, the cost of which is paid by society whether through Medicaid, Emergency Room Care, or insurance policies. These observations provide yet more immediate justification to reduce the constraints on intelligence, thus increasing the intelligence, capability, and employability among the disadvantaged.

Furthermore, as conveyed in Chapter 4, early schooling reduces the number of children labeled Learning Disabled to receive special education services which, incidentally, are disproportionally disadvantaged – an additional problem worth analyzing. In any case, in 2015, Muschkin, Ladd, and Dodge, of Duke University, reported that attendance in two early childhood programs, Smart Start and More-at-Four, reduced the likelihood of a student being placed into special education; "the two programs reduce the odds of special education placement by 39%". Muschkin and her associates add importantly, "that special education services cost 40 percent more than regular class services". Thus, improving nutrition results in health care cost savings, and early schooling results in significantly fewer children placed in special education programs. The combined savings go a long way to pay for early schooling.

The net of these observations is, as poverty decreases so do many costs, most substantially, health care costs. These interventions are a great bargain. Remarkably, they are hardly pursued by society. What would explain that?

---

[128] CDC Health Disparities and Inequalities Report, 2011

Brief review of the relationship between intelligence and school dropout rate, under-employability and other societal problems

Though the data shows an associative relationship between various problems of society and intelligence, applying reasonable and rational analysis demonstrates the relationship is also causal. For example, a well-nourished child is better able to focus in school, is more likely to absorb more information, is more likely to improve intelligence and capability, and is more likely to complete schooling. With that he or she is likely to be more employable, and thus less likely to live in poverty. The general message is clear – raising intelligence substantially decreases school dropout rate, welfare, LBW baby birth rate, criminality, and other problems of society.

The data indicate that most severe societal problems are associated with people with the lowest IQ scores. More specifically,

- 66 percent of school dropouts score in the lowest 20 percent of the IQ spectrum,
- 64 percent of people experiencing under-employability score in the lowest 20 percent of the IQ spectrum,
- 62 percent of people in prison score in the lowest 20 percent of the IQ spectrum,
- 45 percent of mothers with LBW babies score in the lowest 20 percent of the IQ spectrum,
- 63 percent of mothers of children living in poverty score in the lowest 20 percent of the IQ spectrum, and
- 57 percent of welfare recipients score in the lowest 20 percent of the IQ spectrum.

The percent of people directly participating in the societal problems and scoring in the lowest decile (the lowest 10 percent) of the IQ spectrum is much higher than the percent of people directly participating in the problems of society and scoring in the second lowest decile. For example, 42 percent of school dropouts score in the lowest decile of the IQ spectrum and an additional 24 percent of school dropouts score in the second to lowest decile of the IQ spectrum; the curve is very steep in the lower deciles of IQ. Completing the pattern, few people with above average IQ are school dropouts or in prison, or on welfare. Figure 30 shows this graphically. Even the rate of fatality in automobile accidents present a similar curve; the lower the IQ the higher the rate of fatalities in automobile accidents.

As shown earlier, raising the IQ of all those in the lowest quintile of the IQ spectrum by an average of 10 points would increase their IQ to above 87, into the third decile. Results presented in Chapter 6 showed that raising the IQ of the disadvantaged would substantially decrease societal problems such as school dropout rate, chronic welfare services, criminality, and under-employability. It is not possible to predict a precise estimate of the reduction in societal problems resulting from increasing IQ of the disadvantaged but an initial estimate of from 25 percent[129] to 33 percent[130] is justified. For example, as IQ among the disadvantaged is raised by an average of 10 IQ points, the associated school dropout rate is reduced by 30 percent. This is likely an overestimate because there are other influences on dropout rate besides low IQ, but low IQ is the most significant influence. When the evidence presented in Chapter 6 is reviewed in conjunction with awareness that this is not only associative but also causal data, and when this is examined with the understanding that unconstrained intelligence is a powerful tool for the advantaged, the validity will be apparent.

These data apply to health care as well; not surprisingly health improves as nutrition improves. More to the point, health improves as intelligence improves. This results from being more likely to follow medical instructions and more likely to be better informed about health. So, health is improved by both adequate nutrition and, as seen, improved nutrition improves intelligence which, in turn, itself, improves health.

This data strongly supports the fundamental messages discussed in this book; reduce environmental constraints on intelligence and the problems of society, such as both poverty and health, are substantially reduced.

Interventions to reduce poverty versus relief services to reduce the pain of poverty

The solutions discussed in this book do not relate to relief programs for the poor. In fact, the intent of this book is to introduce interventions, approaches

---

[129] $((18 + 20 + 26 + 36)/4)) = 25$; raising the first decile into the second decile
[130] $((30 + 30 + 32 + 42)/4)) = 33.5$; raising both the first and second deciles into the third decile

that reduce poverty, such that relief programs may be reduced – become decreasingly needed; the interventions focus on decreasing environmental constraints on intelligence, thereby increasing intelligence, capability, employability, and thus reducing poverty. To be clear, it is not suggested, here, to immediately reduce relief programs that support the poor, programs such as welfare that reduce the pain of poverty. However, and importantly, it is expected that current relief programs will be naturally and gradually reduced as intelligence, capability, and employability is increased.

Additionally, though we know how to enhance intelligence of most of the disadvantaged, we will need to address motivation as well. There is a mind-set among many that this will be difficult to achieve. We feel that when hope replaces hopelessness, then motivation, optimism, and high expectations will replace frustration.

Underlying these discussions is the expectation that interventions addressed in this book be implemented at the same time, with no interruption, of relief support. *The disadvantaged, especially those in deep poverty, need nutrition, health care, and housing support today – relief solutions. However, given only relief programs, many of those in poverty would likely remain in poverty for decades, even generations. The interventions to enhance intelligence reported in this book result in a substantial reduction in poverty and, with time, not much time, corresponding reduction in candidates requiring relief programs.*

Interventions with relief solutions present a snow-ball affect that needs early and ongoing support. Implementing just interventions, without relief support, would likely prove unethical; there will always be some poor who cannot help themselves. Interventions and relief must be supported in parallel to achieve the desired results, at least initially. This approach is justified because many of the interventions are offered in parallel, quickly take effect, and thus quickly reduce the need for relief support.

Models demonstrating reasonableness of interventions

Are there models that show these, and other interventions really work? Yes! Early school education, mischaracterized as pre-school education, is prevalent in many Western Countries, and is picking up favor in the United States, at least among the advantaged. Providing infants with mother's milk is far more prevalent today than a generation ago though considerably more can be done. Ob Gyn Doctors recommend vitamin and mineral supplements to their prenatal patients, but many disadvantaged pregnant women never see a doctor during pregnancy or do not see one until its later stages. Year-round education is being installed in some places. The importance of effective school lunches in the United States and other parts of the world is becoming increasingly recognized. And, also, parenting training is beginning to be tried but is still in its very earliest stage. These activities have proven their worth, as shown throughout this book. They need to be expanded and strongly supported.

Myths perpetuate the stubborn mystery of intelligence

Research in probably every field of study unintentionally produces myths. The study of intelligence is no exception. Many myths have been introduced some of which have been exposed and debunked. For the study of intelligence to be more productive, many more myths need to be revealed.

One such myth is that intelligence is stable or fixed as some researchers claim. While IQ is stable by design, one's intelligent behavior changes from one moment to the next according to environmental factors encountered, such as nutrition, health, and learning experiences. Adequate nutrition and health allow one's cognitive functions to perform closer to one's peak potential in the moment. Intelligence depends on the performance of one's neuronal system in the moment and is also determined by the effectiveness of processing formal and informal learning experiences. Thus, contrary to widespread opinion, intelligence is essentially open-ended—the more effective the learning experiences, the more knowledge, the more intelligent behavior.

While many claim that intelligence does not include knowledge, this claim is absurd because intelligence surfaces as a behavior, and as such, it must reflect all factors that influence intelligent behavior including knowledge. Surely, if intelligence were characterized only by genetically determined structures such as cognitive structures, essentially pre-determined at conception, then, clearly, intelligence has never been measured. More to the point, knowledge occupies increasing real-estate of the brain as learning experiences are accumulated; they increasingly determine intelligence and intelligent behavior across most, if not all, an individual's life; knowledge growth determines the growth of intelligence. Knowledge has traditionally been omitted from the definition of intelligence because, apparently, hereditarians have influenced the definition; from the hereditarians view only that which is genetically determined may be defined as intelligence. We reject that notion.

Furthermore, it is a myth that differences in intelligence between individuals is *largely* determined by genetic factors; it is a myth that differences in intelligence between groups of individuals are even partly genetically determined; that the disadvantaged are genetically more limited in potential intelligent behavior than the advantaged; that blacks are genetically more limited in potential intelligent behavior than whites; that whites are genetically more limited in potential intelligence than orientals. It is a myth that measurements of human characteristics, in general, such as height, present a normal bell curve distribution of scores; human characteristics are substantially determined by nutrition and other environmental influences, which are not uniformly distributed. Therefore, it is a myth that intelligence-measures should present a normal bell curve distribution, or that IQ tests should be defined to yield a normal bell curve distribution of scores. It is a myth that IQ scores or intelligence measures regress-toward-the-mean. It is a myth that the **g**-factor is one or more **cognitive** genetic factors; **g** is determined by well-being, general well-being, determined by nutrition, health, and other environmental factors; the more well-being, the more one's general capability. **g** should not stand for general intelligence, but general well-being. It is a myth that intelligence does not grow throughout most of one's life; actually, intelligence grows well into the last stages of life. It is a myth that an individual's potential intelligence is limited at conception; potential intelligence is open-ended and grows with every information gained from formal and informal learning experiences including from the now apparently open-ended electronic information bank.

The study of intelligence has been plagued by a myriad of myths. This researcher suggests that a scientifically serious conversation be encouraged to discuss these myths, so that notions that are agreed to be myths are moved into the field's mythology. It is suggested that a conference be convened to finally rid the study of intelligence of these and other myths.

CHAPTER 9
Conclusions, and Directions

The tapestry formed by the interwoven threads of this book may now be viewed in its entirety. Stepping back, the colors and textures of the tapestry merge into a cohesive whole. One can clearly see the interrelationships between the threads of economic inequalities, inequalities of opportunity, poverty, and injustices. Indeed, one can also see the relationships between constrained intelligence and school dropout rate, between intelligence and employability, intelligence and poverty, and intelligence and inadequate or inappropriate nutrition. The tapestry has a unique ability; the threads representing nutrition and health effect the threads representing intelligence; the threads for intelligence get much brighter as the threads for nutrition and health become more vibrant. As the threads for intelligence become brighter so the threads for capabilities, opportunities, and employability, gain strength and durability, which in-turn, causes the threads for poverty as well as economic inequalities to diminish in presence. The threads for motivation, expectations, and orientation become more pronounced as parenting training threads are strengthened. The beauty of the total tapestry is magnified by its scope, modifiability, yet simplicity.

Upon reading this book, a friend asked, "Why have I never heard this rather obvious connection between constrained intelligence and poverty – this circular pattern that, constrained intelligence causes poverty and poverty constrains intelligence." I told him that its much like the rather obvious invention of placing an eraser on a pencil. The observations result from looking at issue(s) with a larger perspective. The issues include upward mobility, unjustifiably large economic inequalities, indefensible inequalities of capabilities, unwarranted inequalities of opportunity, the stubborn intelligence gap between the disadvantaged and advantaged, ways to reduce poverty, approaches that reduce the cost of medical care, ways to reduce welfare, criminality, and other problems of society. These are usually examined individually because of their *apparent* diversity and complexity. If one looks at all these and others at the same time, as we have done in this book, one is told you're trying to boil the ocean; your objectives would be better served by looking at each of these difficult problems separately. In contrast, it seems to me that people, all-too-often, look at a problem through a very narrow lens; look at a sliver of a problem because it is more manageable and more understandable. Here, the observations and interventions result from looking at the problem(s) with a larger perspective, and accordingly, seeing interrelationships and solutions that have not been obvious from an array of narrow viewpoints. Larger views often yield simpler solutions.

So, in this last chapter, while this has been the pattern throughout the book, we *more* closely focus on the bigger picture; we discuss the forest(s), and how they are shaped by their different trees. More basically, we generalize and emphasize the central notion of this book; that there is a *fundamental relationship between intelligence and human activities – **importantly, a relationship that is different depending on whether one is advantaged or disadvantaged;*** that intelligence is environmentally constrained among the poor yet relatively unconstrained among the advantaged; that environmental constraints on intelligence are minimized with adequate nutrition and good health. Human activities form a spectrum. School-completion, employability, creativity, relatively good health, effective formal and informal learning experiences, and self-satisfaction predominate towards one end of the spectrum while welfare services, disproportional medical issues, criminality, injustices, and, of course, poverty predominate towards the other end of the spectrum. The human potential of those near one end of the spectrum is reasonably well used, whereas the human potential of those near the other end of the spectrum is largely wasted.

We know that some problems are better understood and more effectively analyzed by examining a broader perspective, thus potentially recognizing interrelationships that would not have otherwise become evident. An often-told apocryphal story comes to mind of an elephant being analyzed by three researchers each focused on a different part of the elephant, not one scrutinizing with enough information thus each concluding incorrectly but creatively it to be a tree, a fan, or a snake. Analyzing the wider scope of problems often reveals patterns and often suggests the questioning of notions that may have become comfortable and misleading in more narrow analyses.

When viewing a problem with a larger perspective one quickly sees the commonalities to many and different problems, opportunities, and concerns. By reducing environmental constraints on intelligence thereby enhancing intelligence, especially of the poor, one finds that many societal problems may be minimized including many indicated in this book, those not mentioned in this book, and those not yet even perceived. This explains why fundamental notions should, when possible, be viewed from a higher plane.

With a broader view and an understanding of the related evidence, we conclude that significantly reducing, perhaps even largely eliminating, poverty is more readily achieved. By decreasing environmental constraints on intelligence, the poor become more capable, can move out of poverty into the middle-class, and even beyond. More specifically, by 1) providing disadvantaged pregnant women with mineral and vitamin supplements, their babies, at age four, will score an average of 8 points higher on IQ tests, and, separable, their offspring that may have been born as LBW babies will, far more likely, be born as NBW babies and, at age four, score an average of 13 points higher on IQ tests. If, in addition, their infants are 2) provided mother's milk rather than formula, then, at age four, they would score an average of 7 points yet higher on IQ tests.

Examining this increase in IQ more closely is revealing. Enhanced nutrition directly raises IQ in the moment and, indirectly prevents the IQ of the disadvantaged from falling over time. Enhanced nutrition increases an individual's ability to focus in both formal and informal learning settings. It enhances a child's ability to more effectively absorb learning experiences such that subsequent IQ tests show an improved IQ score, or at least a steady IQ score – recall that just maintaining a steady IQ score, after age four, is enough to eliminate the IQ gap between the advantaged and disadvantaged regardless of race. The increased learning likely reduces frustration and increases interest in school. These changes collectively increase the likelihood of staying in school through completion.

Importantly and astoundingly to some, *just these two interventions* *(providing mineral and vitamin supplements to disadvantaged pregnant women raises IQ and also lowers the frequency LBW babies, and providing mother's milk to infants which also raises IQ, as described just above) would go a long way to demonstrate that environmental constraints, typically experienced by the disadvantaged, **explain essentially the entire difference in IQ between the advantaged and disadvantaged at age four, regardless of race**[131]; that while there surely are innate difference in intelligence among individuals, there is **no** justification to assume the existence of innate differences in intelligence between groups of advantaged and disadvantaged people, regardless of race. This central observation, a serious and well substantiated reflection, should stimulate revisiting and reaching closure on this fundamental issue among, at least, the community of psychologists.*

Not so remarkably, evidence provided in Chapters 4 and 5 also shows that as disadvantaged children are provided not only an improved free or price-reduced breakfast and lunch but also mineral and vitamin supplements throughout their developmental years, including school breaks, the gradual and steady decline[132] in IQ, typical of the disadvantaged, is largely prevented; that here too, there is *no* justification to assume the existence of innate differences in intelligence between *groups* of advantaged and disadvantaged people, regardless of race; that with adequate and appropriate nutrition, the disadvantaged absorb new information much like advantaged children, they more likely enjoy school and complete school as do the advantaged, as adults they are more employable and more likely to protect their fruitfulness and that of others, as do the advantaged, and they become members of the middle-class and beyond, as the advantaged, with much fewer medical issues, much less welfare needs, and much more likely to live as content, productive, and contributing members of society. The enormous waste of human potential among most people on the planet will have been largely prevented.

---

[131] See Chapter 5 for details

[132] The IQ gap between blacks and whites gradually and steadily increases by 0.6 points every year between age 4 and 18; that is, though the IQ of the disadvantaged drops by 0.6 points every year of their schooling, the IQ, by design, remains stable among the advantaged. While no genetic explanation has ever surfaced to explain this observation, environmental interventions have been shown that eliminate this effect on the gap.

Surely, other interventions help as well, and many of these would benefit the advantaged as the disadvantaged. These include, parenting training, and early education starting at age three. These interventions would substantially advantage the advantaged. Of course, they would also advantage the disadvantaged.

The pattern – constrained intelligence causes poverty, and poverty constrains intelligence – is not nearly as persistent as may have been thought. Fortunately, this circular pattern consists of links which have severe weaknesses. We have shown that *as environmental constraints are reduced, thereby enhancing intelligence, poverty* in both the developed and developing countries *is greatly reduced;* this stubborn and painful circular pattern can be permanently shattered.

Not only is poverty largely eliminated as the constraints on intelligence are decreased but various societal problems are substantially reduced as well. As their intelligence is increased, disadvantaged children are more likely to complete school and as adults are more likely employable, no longer need welfare, medical problems decrease reducing medical costs, criminality decreases, societal tranquility increases, and GDP increases. Chapter 6 presented details of how interventions, reducing the constraints on intelligence, effect the severity of a wide range of societal problems.

While poverty has many causes, constrained intelligence, typical and most prevalent among the disadvantaged, is the most fundamental, the most widespread, and the most significant explanation of persistent poverty. Evidence, as documented throughout the literature, is clear conveying extensively that the main reason for relatively constrained intelligence among the disadvantaged is inadequate or inappropriate nutrition, poor health, and other environmental constraints. Poor nutrition not only limits the expression of cognitive genetic factors during one's prenatal developmental stage but it also restrains the effectiveness of one's neuronal system during every stage of life. This limits the economically disadvantaged from reaching as close to their potential intelligence as is typically attained among the advantaged. Inadequate nutrition restrains intelligence in the moment and thus limits effectiveness of formal and informal learning experiences, restricting growth of intelligence.

Interventions, reported earlier in this book, break this circular pattern. With these interventions, *environmental constraints on intelligence are*

significantly reduced, the intelligence gap between the economically advantaged
and disadvantaged substantially diminishes – is essentially eliminated, and as
the poor become more intelligent, thus more capable, more opportunities appear;
fewer people live in poverty. As the middle class grows, more goods and services
are demanded, and more profits are made by those providing goods and services.
Additionally, societal costs including for welfare, medical care, under-employ-
ability, and criminality are substantially reduced. And, as shown, interventions
are self-supportive; costs are immediately and entirely offset by savings in related
areas.

These cost savings conveyed in Chapters 1 and 4 explain how the disadvantaged can find and maintain rapid upward mobility *without taking a penny*
from the advantaged. This is true in the United States, the Western countries,
as well as the world over. With this awareness, societies may be more likely
and more willing to establish and maintain the environmental foundation
for everyone to more nearly achieve his or her potential intelligence and
capability. Individuals, having this awareness *themselves*, might be more
willing and likely to strive to attain their own potential.

Several less quantifiable considerations also affect differences in intelligence between the disadvantaged and advantaged such as expectations,
motivations, and orientations.

Certainly, as the disadvantaged move into the middle-class and become
advantaged themselves, the economies of the world grow, perhaps as never
before. However, in recent decades we have seen that nearly all economic
growth ends up in the hands of a tiny minority of the already rich—and
hardly improves the lot of the disadvantaged. Yet while this may be currently
true the now more capable will justifiably expect more appropriate pay,
society will find a way to more equitably share the growing abundance.
This does not suggest a redistribution of wealth—taking from the rich.
Rather, it means sharing the growth; recall that in the 1950s, 60s, and
70s worker's wages in the United States increased at essentially the same
rate as did productivity. From the view of this researcher, again, it is of
little consequence that the rich get richer so long as there are dramatically
fewer poor; that the middle-class increases in number and, accordingly, its
members find and establish their comfortable place in society.

On another concern, it is time that the myths of intelligence be finally
recognized as such. While many fields of study have advanced substantially

over recent centuries, the study of intelligence is an exception, in the extreme. Advances in disciplines such as physics, medicine, genetics, information technologies, and astronomy have dwarfed advances in the study of intelligence, markedly the most distinguishing and most central characteristic of the human condition. What can explain this? Surely, the brain and mind are complex mechanisms. Certainly, experimentation with the brain or mind introduce challenging ethical issues. Though significant advances have been made towards understanding how the brain works, how neurons function, how neurons intercommunicate, and how the neuronal system functions, the workings of the mind remain largely a mystery. Many theories have been proposed regarding how the mind organizes, stores, retrieves, transforms, and interrelates information, as well as how the mind achieves decision-making and creativity, but with little closure. Our explanation, though not complete, is that the study of intelligence has been plagued by *myths* that have dominated the discipline for centuries. It is time that these myths, many described in Chapter 7, be exposed to open-minded debate and either be moved into the fields mythology or shown to be valid explanations that contribute to an understanding of intelligence.

Although poverty has plagued societies for millenniums and has been thought to be a permanent feature of the human condition, we now have reason to believe, we have clear evidence, that *it need not be so*. Surely, there will always be people occupying the lower end of the economic spectrum, but that need not define poverty with pain and hopelessness; the lower end of the spectrum can define the middle-class with upward mobility and opportunity. Society need not be wasteful of human capability whether occupying the upper end or the lower end of the intelligence spectrum.

We have shown that far from being a quixotic quest, if enough people take up this mission and in unison blow at the appropriate windmills, the objectives are well within our reach.

Big ones, indeed

We find it worthwhile to reemphasize and paraphrase a message conveyed by former Israeli President, Shimon Perez, who said that producing small changes takes about the same effort as producing great changes; *go for the big ones*. Here, we surely have focused on big ones – substantially reducing poverty, significantly reducing medical costs especially of the poor, and minimizing various other previously assumed unsolvable societal problems. The approaches described in this book to accomplish this array of objectives do not include redistribution of wealth. Rather, the approaches, here, are to reduce environmental constraints on intelligence such as improving nutrition, thus increasing the intelligence, most substantially of the disadvantaged, and accordingly improving upward mobility, increasing the potential of the disadvantaged by improving their capability, opportunities, and their employability, for the benefit of all in society.

Surely, in the interim, as upward mobility is improved, some of the poor will continue to need help. While overwhelming poverty, economic inequality, and other problems of society have always been with us, this book has shown that society with dramatically reduced poverty can be achieved; that a correspondingly growing and productive middle-class living in varying levels of comfort, is real.

Former US President Clinton once said, "There is nothing wrong in America that can't be fixed by what is right in America." This rather strong assertion presents an interesting challenge, and *importantly*, it likely applies similarly to every group of peoples around the world; there is nothing wrong in Tanzania that can't be fixed by what is right in Tanzania. What is right in Tanzania, the United States, and elsewhere, is the people, and if those in poverty are appropriately appreciated, nourished, and educated, as well as motivated, they will likely find their way out of poverty.

Many people in the developed world, and many more in the underdeveloped world, have been constrained from reaching their potential intelligence and capability resulting in much unnecessary poverty, pain, and suffering. Importantly, *this is not to say they cannot achieve. This is to say, they have not achieved.* More specifically, it is to say, as nutritional and other environmental constraints are decreased, many people would likely more closely reach their potential intelligent behavior and become more productive members of society – it is to say that much of their potential capability and potential intelligence has been wasted and should not, need not, continue to be wasted.

On a related concern, why is there an IQ gap between the disadvantaged and the advantaged? This was discussed in Chapters 4 and 5. Note the question is not *whether* but *why* there is this IQ gap. Similarly, why is there an IQ gap between the blacks[133] and whites? The answer to both these questions has been shown to be the same. While there is no evidence to support the claim that genetic factors cause the IQ gap between groups of people, there is substantial evidence that IQ gaps between groups of people are caused by environmental constraints resulting from being economically disadvantaged. Recall that blacks, in the United States, are three times more likely disadvantaged than whites. The answers to both the above questions have been shown to be environmental constraints due to living in poverty or near poverty and thus experiencing, for example, inadequate nutrition as well as limited health, and, accordingly, less than effective formal and informal learning experiences. Interestingly, the answers to these same questions, focusing on opportunity inequalities and economic inequalities, are also the same. *The explanation for the IQ gaps is not race but economic disadvantage.*

As shown in this book, building on widely accepted literature, intelligence, among the disadvantaged, is significantly constrained by environmental factors, while, among the advantaged, intelligence is largely supported by environmental factors. Looking at this observation as a continuum, all of us experience varying levels of environmental constraints preventing us from reaching our potential; none of us ever reaches our potential intelligent behavior. *No one!* The disadvantaged of the world experience much more environmental constraints and thus are more severely prevented from reaching their potential. Some of us get close, while some hardly reach a level of basic functioning. The disadvantaged in the developed world come significantly short of their potential intelligent behavior. The disadvantaged who live in the developing world come far short of their potential intelligent behavior. Often their potential capability is hardly tapped, and their self-image and the image others have of them is, itself, limiting. The extraordinary potential of one's human capability, permitted, for example, by one's intelligent behavior, is severely limited for billions of people by environmental constraints, which, again, importantly, *can be reduced.*

Opportunity is a core requirement for each one of us to come reasonably close to achieving our potential capability. Reducing environmental constraints so that each individual may more closely reach his or her potential

---

[133] The distinction between being black or white has always been less than well defined. The usually accepted definition is self-definition.

intelligent behavior, and thereby be open to greater opportunities, is a central message of this book. As shown, enhancing intelligent behavior to permit each individual to more closely reach his or her potential, especially and most readily of the billions of disadvantaged, significantly reduces poverty, diminishes economic inequalities, and decreases many societal problems.

The notion that people are *created equal*, mentioned in the United States Declaration of Independence, possibly first documented by John Locke, paraphrased by Abraham Lincoln in his Gettysburg Address, has now become a central theme in many societies. But what does *created equal* mean? Does it refer to *equality under the law*, as some have decided? This is first indicated in the Fourteenth Amendment of the United States Constitution. That interpretation has seen much progress. Does it mean all people should have *equal opportunity,* as some may believe? We are all unique – different in many ways. We all live in different times, in different places, with different objectives, with different talents, capabilities, and interests. So, it appears inappropriate to even strive to achieve *equality* of opportunity. Likewise, it is inappropriate to strive to achieve economic *equality.* However, and *this is central, we can minimize inequalities – economic inequalities, and opportunity inequalities.* We can minimize environmental constraints, over which we have significant control; we can minimize nutritional and health constraints. We have seen that by reducing environmental constraints, constraints on opportunity would also be reduced, the economic gap between individuals, and between groups of individuals, would be reduced; each individual would, likely, more nearly reach his or her potential.

While poverty certainly may result from many causes, here, we focus on the effects of limited intelligence caused by environmental constraints. Those in poverty raise their children in poverty who, because of limited resources, will, in turn, likely experience limited nutrition and poor health and thus limited intelligent behavior as well. Many break out of the cycle, but many more do not. As mentioned, in the United States 45 percent do not break out of poverty – not something of which to be proud. In many other developed countries, the percentages are better, but in much of the world, without reducing environmental constraints, the odds convey that it is nearly impossible.

We know how to significantly enhance intelligent behavior, especially of the disadvantaged, at no cost to society. That results in helping hundreds of millions in the developed world and billions in the developing world to come much closer to their potential intelligent behavior. This permits each person to more closely reach his or her potential opportunity – to reach a higher level of self-satisfaction, and to help vast numbers of people to more likely experience human dignity. It would go a long way to reducing under-employability and other problems of society. It would benefit not only those directly involved with these societal problems, the disadvantaged, but those indirectly involved with these societal difficulties, the advantaged. With just the interventions shown in Chapters 4 and 5, the IQ gap between the advantaged and disadvantaged, between blacks and whites, can be essentially eliminated.

Surely, intelligent behavior and upward mobility are directly related; the more intelligence the more capability, the more upward mobility. Both intelligent behavior and upward mobility are inversely related to problems of society, including poverty, economic inequalities, and opportunity inequalities; the more intelligent behavior and upward mobility, the fewer problems of society, the less poverty, the less economic inequalities, and the less opportunity inequalities.

What's more, with increased intelligent behavior, especially women's intelligent behavior and women's education, population growth-rate decreases and then levels off at a lower level, and this alone would change the dynamic of the concerns mentioned. With a more stable population size, society can spend more resources on education, sanitation, health, clean water, and energy. Additionally, women will be more available for, more able and more willing to enter, the job market, further reducing poverty. Also, with enhanced intelligent behavior there will be more and better educated people addressing and likely solving more of the severe existential problems of society. Enhancing intelligent behavior is not the only answer but it is likely the single most effective answer.

Though these results are significant, they only scratch the surface. The real benefits are realized by each individual in society; among the disadvantaged as increased self-confidence, self-reliance, and increased hopefulness for themselves and their children, and among the advantaged by, at least, better education for their children, greater societal stability and increased income. And again, the interventions described are self-supportive; they do not take from the rich to give to the poor. Really!!

Economic alternatives claimed to reduce poverty and economic inequalities

To reduce poverty and reduce economic inequalities, a wide range of solutions have been offered over the years by economists including altering the distribution of wealth[134]. However, the benefits of these purely economic approaches may, perhaps, only reach a relatively few in poverty, and the effects may well be short-lived because they hardly address the underlying issues of poverty directly; barely touching the need to increase the intelligence and capability of the disadvantaged.

Some sociologists believe that economic inequalities are a natural phenomenon that will only get worse. Charles Murray, a political scientist, points out in his 2012 book that the advantaged attend the best schools, find their mates in the best schools, have children who are brought up with the most advantages within gated communities, who, throughout their childhood get prepared for the best schools, and the cycle continues with the gated communities perhaps becoming increasingly select. Murray's projection is of a society ripe for conflict and disruption.

---

[134] In his recent white paper, Stiglitz suggests changing the tax structure to keep corporate profits in the United States so that more jobs may be created in the US, more infrastructure work be established, and more monies be used here to reduce deficits and debt. In 2014, Piketty's book, *Capital*, suggests that though we cannot predict the future, we can gain a better understanding of the future by studying the past. He disagrees with Friedman's 1960s views of monetary policies—Friedman's view that we don't need a welfare state, or Friedman's view that we do not need progressive taxation. Piketty fosters Stiglitz's fiscal changes. He suggests such changes would reduce economic inequalities and reduce and then stabilize economic inequalities at more reasonable levels. Though Piketty's central thrust deals with the role of capital in wealth creation, he favors more effective democracy and more democracy transparency, as well as, financial transparency. Incidentally, Piketty encourages, as does this author, and many others, that more education and more early quality education, starting at age three would reduce economic inequalities.
Robert Reich, in 2014, advocated for a few more proposals such as "limit the size of banks, make it easier for low-wage workers to unionize, [and] raise taxes on corporations with high ratios of CEO-pay to average worker-pay."
A more recent book entitled, *Inequality: What can be done*, written by Anthony Atkinson, favors yet a different approach to reduce economic inequality. According to the *Economist*, June 6th, 2015, "he dwells on one class of contributory factors above all others: the subtle (and not-so-subtle) ways the rich are able to influence governmental policy in order to protect their wealth."

This book conveys essentially the opposite projection—it conveys the potential of an integrated society; that by enhancing intelligence, especially of the disadvantaged, most of the poor will have the opportunity to move out of poverty into the middle-class and beyond. With the evidence reported in this book, this author disagrees with Murray's claim that the intelligence gap between the cognitive elites and the rest of the population will only increase over time. Murray's conclusion is based on an unfounded assumption—the myth that differences in intelligence are largely determined by genetic factors, also a basic assumption of his earlier book, *The Bell Curve*, published in 1994, coauthored with Richard Herrnstein.

Though some economic approaches to reduce economic inequalities have their merits, *this book has not advocated any of these;* this book conveys an entirely different approach. Economic approaches such as those suggested, just above, often provide indirect benefits for the poor. However, it is argued that the approaches described here, of enhancing intelligence and thus capability and thus opportunity, have a far more direct and lasting effect in helping the poor move out of poverty. The approaches described in this book enhance employability *and* raise salaries. They raise self-confidence, self-reliance, and entrepreneurship, thus accelerating upward mobility. This is not to suggest one set of approaches over another. It is to suggest that while various approaches have merits, have differing objectives, they may be pursued in parallel.

Some members of the general-public, some public policy makers, wanabe economists, might proclaim, "What will these additional better educated people do? They will only compete for the already existing jobs." No! they will not take jobs of others, they will create jobs. Commented earlier, as the middle-class increases, demand for goods and services increases, thus, there will be more jobs to be had, jobs in response to the greater demand.

With more people having increased intelligence and capability, there will likely be increased innovation addressing humanity's existential problems[135], perhaps directed at climate change and alternative energy creation,

---

[135] Some may observe that as the middle-class grows so does demand for energy which, in turn, may result in increased climate change. On the other hand, as the number of advantaged people increases—and we would consider the middle-class among the advantaged—population growth likely decreases, especially if women have equal educational opportunities. However, even as population growth decreases, populations will

and this will likely result in profound changes in human relations among the various peoples, with dramatic implications on the stability of life on the planet. While well beyond the scope of this book, it would be inappropriate to ignore these implications. Enhancing the intelligence and capability of billions of people will result in profound changes in humanity's productivity and creativity.

Three-legged stool

Achieving enhanced intelligence and capability of each individual, especially of the disadvantaged, requires the realization of three related objectives, each of them a leg of a figurative three-legged stool. These objectives are,

a. Individuals in society, whether disadvantaged or advantaged, need to understand and accept the notion that environmental constraints can be reduced thus improving intelligence, most substantially of the disadvantaged, thus enhancing capability and employability.

b. Society needs to provide both the practical and legal environments necessary to permit *individuals* to more closely reach their potential intelligence. These include educational, judicial, economic, medical, and social constructs needed for all to more closely reach their potential. This requires equality under the law limiting prejudices and biases. And,

---

likely increase, albeit at a lower rate. Furthermore, as more people become more capable and better educated, solutions to energy issues may well be increasingly introduced. As innovation regarding alternative (non-fossil-fuel) energy creation is introduced, the price of alternative energy is likely to decrease. More interestingly, as more effort is devoted to developing alternative energy, with a focused objective to reduce global warming, perhaps, accordingly, to make fossil-fuel use obsolete, concern may develop that alternative energy availability might gradually reduce and eventually obviate the need for fossil fuels. (Note that the need for fossil-fuels may well continue in the development of plastics and many other products.) With the fear that alternative, non-fossil-fuel, and cheaper energy, may become available, current fossil-fuel energy companies and countries may decrease prices and therefore affect the well-being of countries and populations currently dependent on fossil-fuel energy production such as countries in the Middle-East, Russia, and Venezuela. These populations will come into tough times (some have already) affecting geopolitical relationships.

c. Both the disadvantaged and the advantaged are likely more successful when driven by high expectations, motivation, and encouragement; so that each has the inner drive and interest to strive toward
his or her potential capability and potential intelligence.

Individual's Capability

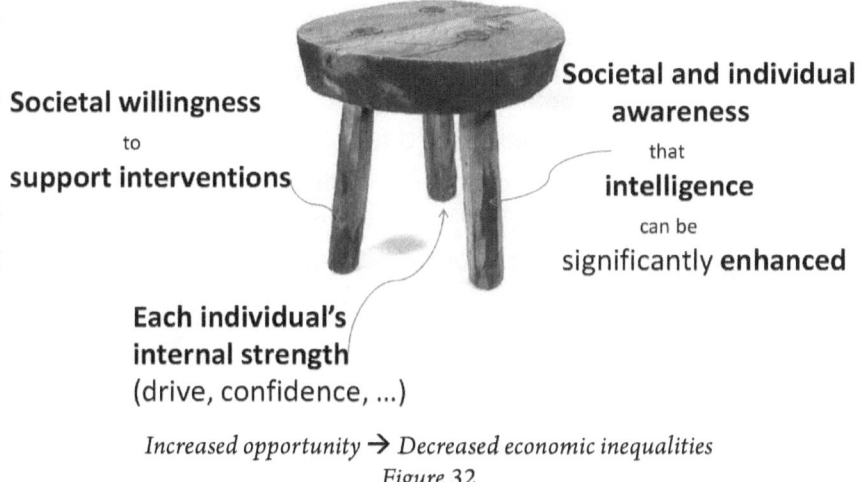

Societal willingness
to
support interventions

Societal and individual
awareness
that
intelligence
can be
significantly **enhanced**

Each individual's
internal strength
(drive, confidence, ...)

Increased opportunity → *Decreased economic inequalities*
Figure 32

The stool is most useful with all three legs being strong and functional;
any one or two legs are insufficient to reliably support the stool; the stability of the stool depends on the degree to which all three legs are firmly
positioned.

These three sets of objectives have been discussed in this book with
varying intensity mostly conveying the awareness that intelligence can be
substantially improved through enhanced nutrition and other environmental factors. Here, additionally, we emphasize society's responsibility to assure the availability of the needed environments, and we also emphasize
everyone's responsibility to create, perhaps with help, the drive and interest
to more closely reach his or her own potential.

Societal constructs, individual internal strengths, and the awareness
that intelligence develops and may be enhanced across one's life, are the
three legs of a stool on which individual potential rests. If society provides

only limited educational, health, and other societal structures, then individual potential is limited. If intelligence is believed to be essentially fixed, predetermined, and unchanging throughout life, then individual potential is stymied. If one's internal-individual strength, one's self-confidence and motivation are weak, then individual potential is stunted. The stability of these legs is not entirely within the individual's control. Suggested are the aggregate of collective interventions to achieve enhanced intelligence to more closely reach one's potential.

In sum, the first leg reflects the need to provide the advantaged and disadvantaged the knowledge and assurance that especially the poor can get much closer to their potential intelligence and capability. The second leg conveys that judicial, medical, and other societal structures be in place to provide the underpinnings for each individual to achieve improved intelligence. And the third leg conveys that self-confidence, motivation, and high expectations are needed to provide individuals with the strength and courage to get closer to his or her potential intelligence and capability.

Expanding on the second leg: society's responsibility

As reported in a recent poll, Kristoff "found that 71 percent of voters supported a major federal investment in early education, including huge majorities of Democrats, Republicans and independents."

Yet, as mentioned, only 38 percent of three-year-old children in the United States, probably advantaged, receive formal education whereas 70 percent receive such education in the rest of the Western world. Early education, along with prenatal care and parenting training with an underlying well-being including adequate nutrition and health throughout the developmental stages, hold great promise to enhance intelligence, increase employability, increase opportunities, and decrease poverty as well as economic inequalities. This would not only benefit the disadvantaged but the advantaged and the very rich as well.

But, I would add, early education is only one of the suggested interventions. While this one is of significance, it along with low-hanging fruit would significantly enhance intelligence especially of the disadvantaged.

On a more general concern, Andrew Cuomo, three term Governor of New York State, tells this story.

When I see a society that looks at a child who is Black or Latino or whatever and poor and concludes all kinds of stereotypical things like, there's no point in trying to help them, they're not as smart as we are, when I start reading books by presumably intelligent people, which are being read by intelligent business men, who ask me at a meeting, "what about this study that says that Blacks are intellectually inferior", and I hear that in New York State, and my only answer to the guy is, are you kidding? Do you believe it? Is it serious? Do you really think that there is something to the thesis that because they are Black they are inferior in intelligence? Do you think that is worthy of discussion? Absolutely, says this White business man.

Now in 2018, while we know much has changed, we also know much has not changed. Change regarding these issues requires widespread knowledge, awareness, and emotional acceptance of the basic notion that, *regardless of race*, intelligence may be significantly enhanced, most substantially of the disadvantaged, can increase capability, employability, and thus dramatically decrease poverty. This book focuses on conveying this awareness.

The third leg: Inner drive to strive toward potential intelligence and capability

The day of a test, one girl told me, "I had a good breakfast, as you said I should, but my mother told me, 'It's no use. You are stupid.'" This kind of messaging, especially from one's mother, likely creates a lasting self-perception and may be devastating. The reverse form of messaging is also often experienced. Carol Dweck, of Stanford University, has reported that telling a child that she is bright may be less effective than telling her "I am proud of you. That took hard work."

Dweck, has studied and written extensively about perception of intelligence. She concludes that one's perception of intelligence determines whether and how we strive to reach our potential for ourselves and our children. If intelligence is believed to be stable, that it is determined at conception, then with failure one is likely to conclude, "That's the best I can do". However, if intelligence is believed to dynamically change then with failure one is likely to say, "I must have examined the problems the wrong way. I must try harder next time." In the first case, where one believes intelligence is established at conception and stable, one assumes fate determines intelligence and capability; there is not much one can do about it. In the second case, where one believes intelligence changes with environmental conditions, one assumes that one has considerable control over one's intelligence and capability. People need to be aware that they do have significant control over their own and their children's intelligence.

Dweck's findings would have far greater impact if one's perception of intelligence were more closely related to a clear understanding of intelligence—that intelligence is not fixed but that irrefutable evidence indicates that intelligence is dynamically modifiable with every formal and informal learning experience, every change in the environment, adequate nutrition, and well-being. Research concerning intelligence would be far more productive if the environmental influences on intelligence were fully acknowledged. Educational resources might well be differently provided if it were recognized that an individual's intelligence is not stable but changes throughout life with every environmental experience.

This was highlighted, over a century ago by Alfred Binet, the developer of the first IQ test. The perception of intelligence in school systems appears to encourage educators to test students and, using the results, predict each individual's potential. It is worth repeating that while one can predict the potential of a group of individuals according to their average IQ score, one can hardly predict the potential of any one individual in the group. A more accurate perception of intelligence is vital to the success of the entire educational process.

An awareness that intelligence constantly changes affects every facet of our existence, our relations with others, and our perception of our own potential. While this book focuses on conveying an awareness that intelligence is largely determined by one's environment, and the associated understanding that an individual's intelligence constantly changes, this book also addresses societal responsibilities to help improve the intelligence, especially of disadvantaged individuals.

Hopelessness, typical of poverty, is not a necessary feature of the vast majority of humans. Hopelessness may be reversed and transformed into hopefulness by enhancing intelligence and thus increasing the likelihood of successful experiences. With hopefulness, poverty is likely reduced and perhaps even largely eliminated.

The individual's and society's basic understanding of intelligence determines not only how we deal with our own potential, but also how society deals with those in poverty, the unemployed, school dropouts, the prison population, and many other societal considerations. Understanding the facts that intelligence constantly changes gives support to a belief that one's intelligence may be improved. This awareness provides the foundation for each one of us to strive more effectively towards our potential and, accordingly, our potential place in society. It allows society to reduce poverty, unemployment, and prison populations by enhancing the lot of the disadvantaged as well as that of the advantaged. It permits capabilities to be expanded and opportunities to be revealed.

Misunderstanding intelligence results in biases and injustices

Most of the world's population, whether white, black, Asian, or Native Americans, is economically disadvantaged and would benefit from reduced constraints on intelligence – enhanced environmental conditions including improved nutrition and better health. These constraints also result from societal conditions and often societal biases, biases against other groups and even, importantly, biases of individuals against their own group. A story, perhaps legendary, is told that Gandhi was once asked, "*What do you think of Western civilization?*" He supposedly responded, "*I think it would be a good idea*". Gandhi, of course, thought Western civilization came short of being a civilized society in its treatment of people, certainly the people of India. Today the West treats many, even its own people, not much better. And while a great deal has changed during the last century, much remains wrong with Western society and other societies the world over. Nearly half the people in the Western world are disadvantaged, and the percentages are far worse in the rest of the world.

The subtle and sometimes not-so-subtle complacency to the widespread pain and quite suffering of the poor is evident in essentially all societies. Injustices caused by biases, or prejudices, or simply by deeply ingrained false stereotypes, have a way of preventing human potential to surface. Of course, this is also true for the disabled. *One child restricted to a wheel-chair was being wheeled to his next class by his mother who, upon coming across one of his teachers, gleefully said that her son had gotten straight 'A's on his recent report card. The teacher responded to the mother, in front of the son sitting in the wheel-chair, "What's the big deal? What else can he do?"* People can be cruel, sometimes unknowingly. Societies can be cruel as well.

Injustices take several forms. I tutored and mentored a twelve-year-old boy whose mother, a bright, and sensitive woman, was 13 years older than he. To my great surprise, this frustrated and troubled child, we'll call Bob, asked me to work with him and though I hesitated at first, two days later upon arriving at school, handed me a note from his mother giving me permission to work with him. This time I did not hesitate. Within about five or six weeks, Bob moved from being a straight F student, and on-the-surface proud of it, to be a successful student, beginning to get 'A's in classes. Let me quickly clarify. My tutoring, alone, did not cause the change. My listening, caring, and motivating helped cause the change. He needed high expectations. He needed to feel that he could meet high expectations.

Several weeks after starting to mentor/tutor Bob, his mother asked me to attend a school meeting about her son. (School systems across the US are required to have a meeting once a year with parents, teachers, and the guidance counselor for every child labeled learning-disabled.) For the first ten or fifteen minutes of the meeting, the teachers described Bob's poor performance during most of the year. He was disruptive, difficult, and unmanageable. When they completed their comments, and knowing his IQ score was 115, I said that, according to his IQ score, he should be among the three or four highest performers in the class; I had reason to believe he was not learning-disabled but disabled by the baggage weighing him down. Upon reminding the teachers that he had recently gotten several 'A's in various classes, I thanked the math teacher for causing her class to give him a standing ovation; precisely what he needed. I asked the members of the meeting whether they had noticed his handwriting—the most artful and creative I had ever seen—whether they had seen his doodles when he was bored in class—this child had a strong artistic bent, and I asked whether anyone knew that Bob played keyboard music in church every week-end. I turned to the mother and said "These people are aware of only one side of your son; they hardly know your son. You need to let them know; you need to advocate for him." They wanted to keep him in resource classes (remedial classes for learning-disabled children) until he proved he could do the work. I suggested that he be placed in regular classes until he proved that he could not do the work. I suggested, additionally, that he be placed in classes for art and or music to help him find real satisfaction in school. Bob's mother nodded agreement. Changes were gradually made which proved to make a big difference.

This raises a few basic questions. Is the United States learning-disabled program working as originally planned? Why are blacks disproportionately labeled learning-disabled? And why is it that "Seventy percent of all school-age children who are diagnosed with learning disabilities are boys."[136] On a more personal note, why have most of the children I tutored and mentored been labeled learning-disabled, when, often, their problem was really the baggage they carried? Why have so many children told me, "I am black. I am stupid." This perception results from society's view but also from blacks talking of themselves, to their children. How has society influenced blacks to have this view of themselves? Why has it been widely assumed, in contrast to the data conveyed in this book, that blacks have lower IQs because of cognitive genetic limitations rather than because of economic disadvantages? These questions reflect misinformed, deeply ingrained, societal biases. And similar questions may be asked about the disadvantaged in general.

Highlighting racial injustices, Flynn wrote, "In 2008, African Americans face a mainstream white culture which tells them they have "worse" genes for IQ than white Americans" and that "'irresponsible' sexual behaviour dooms well over half of their children to live in single-parent homes and poverty."[137] While, without doubt, blacks currently have lower IQs than whites, it is not because of "worse" genes but because of more environmental constraints on the development of their intelligence. And while a large

---

[136] *Newsweek*, September 19, 2005, p. 59.
[137] See more at: http://www.cambridgeblog.org/2008/09/james-flynn-in-the-new-scientist/#comment-87614639

301

percentage of "black American children will grow up in a family with just one parent, normally a woman", an increasing percentage of white children will, likewise, grow up in a family with just a mother. An array of complex reasons explains this trend for both whites and blacks.

Furthermore, as reported by the NAACP in 2014, five times as many whites are using illicit drugs as African Americans, yet African Americans are sent to prison for drug offenses at ten times the rate of whites. African Americans represent 12 percent of the total population of drug users, but 38 percent of those arrested for drug offenses, and 59 percent of those in state prison for a drug offense. African Americans serve virtually as much time in prison for a drug offense (58.7 months) as whites do for a violent offense (61.7 months). Martin Luther King said, "One hundred years later [after the Civil War] the negro lives on a lonely island of poverty in a vast ocean of prosperity", and that, said in 1963, more than a half century ago, is still true today. Why is this problem so stubborn and persistent? What could explain this problem and these biases against millions in the United States, alone? This data strongly suggests entrenched prejudices resulting in severe injustice.

Of course, these questions and data suggest prejudice against blacks. But blacks do not have a monopoly receiving prejudice; blacks, themselves, are prejudiced. Prejudice and biases take many forms and applies across races, religions, nationalities, and economic status, in all directions. So, I ask, why? What are the underlying causes for bias? What are the effects of this prejudice, the world over? Why did the world stand by during the machete genocide of 900,000 people that took place in Rwanda? Why did the world watch while 2½ million Cambodians were slaughtered, while a million Armenians were marched to their death, while 2 million Ukrainians were starved to death, while six million Jews were gassed to death, and while many millions of Native Americans were killed? Answers are varied, glib, and shallow.

Conveying a specific example, during the Holocaust about 25,000[138] people offered help to save Jews from death with great danger to themselves and their families. Later, the Israeli government awarded them by naming them *Righteous Among the Nations*. Surely, 25,000 is a huge number of good

---

[138] As of January 1, 2014, Yad Vashem recognized 25,271 Righteous Among the Nations from 49 countries.

and courageous people. However, placing this in perspective, and realizing there were more than 200 million people in Europe at the time, this represents about one person in eight thousand. The other 7,999 in eight thousand people were either supporters, directly or indirectly, of the genocide, or simply *indifferent*. Surely, some, but very few, were unaware.

Speaking, now, more generally, Abraham Joshua Heschel said, "The opposite of good is not evil. The opposite of good is indifference." Similarly, Edmund Burke said, "All that is necessary for evil to prevail is for good people to do nothing."

A person who is indifferent to injustices is arguably nearly as guilty as one who is directly involved with the injustices. That may seem too harsh, but I contend that indifference empowers the guilty and gives them license and leeway to do as they please. In this book, however, we are addressing a different kind of injustice—the languishing of billions of disadvantaged people in the world, many severely economically disadvantaged. Most of these injustices can be significantly reduced, but only with the active interest of those who would otherwise be indifferent. All of us are busy with important activities, but that only rationalizes indifference.

Isn't it interesting that there is abundant data comparing the IQ of blacks and whites the world over yet there is much less comparable data comparing the intelligence of the disadvantaged and the advantaged? Why do so many people, aware of the severely limited lives of billions of disadvantaged, most of whom destitute, hardly give them a passing thought? Why is there so much complacency to their plight? Indifference to pain and suffering is not new. Injustices are not new. Prejudice is not new. The magnitude of these prejudices and injustices *is* new, *appears to be* growing in some ways, yet, importantly, the means to reverse these injustices *is well within our reach.*

Of course, economic injustices existed thousands of years ago, in every economic and political system. In the United States, as recently as generations ago, child labor was common, unsafe working conditions were hardly questioned, low wages were widely accepted, and economic insecurity was the rule. While much has changed, injustices *benefitting corporate entities* still exist—witness the fires in the garment industry in Bangladesh, chemical factory explosion in India, coal-ash spills in North Carolina, coal mine disasters in West Virginia, ignition and other safety problems in the automobile industry, and oil spills on the Deep-Water Horizon Rig in the Gulf of Mexico and Exxon Valdez Oil spill. The basics of human nature often drive people to optimize *my* objectives without regard to that of others by, for example, minimizing wages, reducing competition by forming a monopoly, or ignoring environmental concerns. Surely, while these injustices have diminished in number, there remains a mindset—though certainly not universal—that seems oriented to optimizing corporate profits at the expense of the environment, employee overall welfare, and the general well-being of society.

A powerful example of protecting corporate growth and profits at the expense of the health of the poor and, in fact, the advantaged as well, was reported in the NY Times on July 8, 2018. They wrote,

> A resolution to encourage breast-feeding was expected to be approved quickly and easily by the hundreds of government delegates who gathered this spring in Geneva for the United Nations-affiliated World Health Assembly. Based on decades of research, the resolution says that mother's milk is healthiest for children and countries should strive to limit the inaccurate or misleading marketing of breast milk substitutes.
>
> Then the United States delegation, embracing the interests of infant formula manufacturers, upended the deliberations. American officials … turned to threats, according to diplomats and government officials who took part in the discussions. Ecuador, which had planned to introduce the measure, was the first to find itself in the cross hairs. The Americans were blunt: If refused to drop the resolution, Washington would unleash punishing trade measures and withdraw crucial military aid. The Ecuadorean government quickly acquiesced.
>
> It was the Russians who ultimately stepped in to introduce the measure — and the Americans did not threaten them. … A Russian delegate said … "we feel that it is wrong when a big country tries to push around some very small countries, especially on an issue that is really important for the rest of the world,"

The Times references two articles published in the journal *Lancet*[139]. The first, written by Victora and associates[140], conveys the importance of breastfeeding. They write, breastfeeding "could prevent 832,000 thousand annual deaths in children younger than five years and 20,000 annual deaths from breast cancer" worldwide. As huge as this appears, it is only a tiny portion of the tip of the iceberg. Additional tens of millions who, because of being provided limited mother's milk, "have lower infectious morbidity", thus live in relatively poor health and limited capability.

The second article, written by Rollins and associates, addresses ways to encourage breastfeeding. They indicate, for example, that not "breastfeeding is associated with lower intelligence and economic losses of about $302 billion annually". Yet, earlier this year US delegates at the World Health Organization, WHO, tried to prevent fostering mother's milk over formula apparently to protect the all-important revenues, more than $40 billion, of the formula suppliers.

If this were an isolated incident then perhaps one could say this was misinterpreted but according to the same NY Times article, there is more. Specifically, at the same World Health Assembly "the United States succeeded in removing statements supporting soda taxes from a document that advises countries grappling with soaring rates of obesity." Additionally, "Washington, supporting the pharmaceutical industry, has long resisted calls to modify patent laws as a way of increasing drug availability in the developing world."

Clearly, there are forces in place that would tend to support the objectives of big business; forces that would suggest the poor improve their own health and get themselves out of poverty; forces that tend to minimize the support given the poor. But why would people prevent the implementation of the interventions discussed in this book that cost society nothing? Why would people prevent increasing intelligence and capability, especially of the disadvantaged, thus substantially reducing poverty, medical costs, criminality, and welfare while, at the same time, increasing the demand for goods and services? The objectives of this book are to raise awareness that supporting big business without improving the lot of the poor leaves huge

---

[139] Victora and associates, and Rollins and associates
[140] Victora and associates (2015)

profits on the table. Many still hold deeply ingrained beliefs that helping the poor out of poverty is useless and hopeless. While that view may have had some credibility in the past, it no longer has any standing, given the evidence provided in this book.

Returning to the main flow, while *eliminating* economic inequalities is *not* a meaningful objective, reducing economic inequalities so that poverty is dramatically reduced and so that the middle-class is correspondingly increased in size is not only achievable but would significantly benefit all, the disadvantaged as well as the advantaged, including the very richest in society. And, as shown in this book, this may be accomplished *not* by taking from the rich but by enhancing the capabilities and opportunities of all members of society. As former US Secretary of Labor, Robert Reich, said, the answer is not simply to raise taxes on the rich and redistribute funds to everyone else, but rather we need an agenda for shared prosperity. This author would add, we need an agenda to raise the capability and opportunity of the disadvantaged. This book offers such an agenda without taking a penny from the rich—just the opposite, providing significant benefits to the rich.

Conclusions and directions

The single most important cause of poverty is environmentally constrained intelligence resulting from inadequate nutrition, poor health, and thus, less than effective formal and informal learning experiences; in short, a major cause of poverty is poverty itself. It is not race, but economic disadvantage. It is not genetic factors, but environment factors. It is not nature, but nurture. Poverty is also likely partly due to a lack of inner-drive that stems from hopelessness, one of poverty's side effects. Poverty all-too-often creates a circular pattern, one that, fortunately, can be broken.

As measured by IQ tests, the advantaged behave more intelligently, than do the disadvantaged; by increasing the intelligence of all of us, most readily and abundantly of the disadvantaged, we significantly reduce economic inequalities, opportunity inequalities, and poverty.

It has been shown that intelligence-inequalities are a major cause of economic inequalities and opportunity inequalities. Though, of course, it is not rational to try to eliminate any of these inequalities, it is reasonable and beneficial to significantly reduce these inequalities. Interestingly and importantly, discussing this set of issues from the viewpoint of intelligence-inequalities makes clear that we are not suggesting taking from the rich to give to the poor; one cannot take intelligence from one to increase the intelligence of another.

We have reported several environmental interventions that would substantially increase intelligence, especially of the disadvantaged. An objective of this book has been to demonstrate the importance of enhancing intelligence so that the upward mobility path is more available; so that those in poverty may more closely reach their potential, and thus reduce many of society's most persistent and painful issues. Specifically, we focused on a collection of interventions to reduce environmental constraints on intelligence – to enhance intelligence – not only to decrease the school dropout rate and increase employability, but to reduce poverty and the associated hopelessness.

Some might argue that interventions require societal financial investment. But this book has shown that many interventions produce immediate results, and, that in fact, the short-term benefits, savings, substantially exceed the costs, and the long-term benefits far exceed the costs. As shown, reducing the frequency of LBW baby births is only one example, one medical issue disproportionately experienced by the poor, whose immediate savings and whose long-term benefits far exceed its cost.

Another circular pattern is apparent. Not only does nutrition effect intelligence and intelligence effect nutrition, but health effects intelligence and intelligence effects health. As health improves it is likely that intelligence improves. And as intelligence improves it is likely that health improves as well. These observations have huge implications on society and on how society justifies the expenditure of limited resources. Health-care costs could be substantially reduced by raising intelligence, especially of those near the low end of the intelligence spectrum. The cost of raising intelligence should prove trivial when compared to the resulting cost savings of maintaining good health. This represents yet another justification to reduce environmental constraints on intelligence.

Problems discussed in this book are many, varied, and complex, and are experienced in Western countries as well as across the world. It may seem absurd to claim that a single approach would solve much of those problems. However, we have conveyed just that; that enhancing intelligence, especially, most substantially, and most readily among the disadvantaged, would go a long way to resolve many problems of society. Enhanced intelligent behavior would increase prospects, especially among the disadvantaged, and thus reduce the opportunities gap. It would significantly reduce school dropout rate, under-employability, criminality, and many other societal problems.

Furthermore, it has been shown that as intelligence is enhanced, especially of the disadvantaged, poverty likely significantly declines the world over; there is more ownership of society's well-being, more pride in oneself, in one's potential, and in that of one's children; *enhancing upward mobility and increasing the middle-class likely increases one's stake in society and its well-being benefitting all directly and indirectly.*

We have shown the IQ gap between blacks and whites is environmentally determined. Blacks, in the United States disproportionately encounter far more environmental constraints including inadequate nutrition, poor health, and thus, less than effective formal and informal learning experiences limiting the growth of intelligence especially during the many developmental stages of life. There is an IQ gap between other groups of people as well, for example, between Asians and whites. Here also, while no genetic explanations have ever surfaced, environmental differences explain the IQ differences. More generally, *the IQ gap between racially or culturally different groups of people is essentially due to environmental constraints.* This significant observation is central to understanding the causes of severe economic inequalities and is vital to improving the well-being of many people, billions who have had difficulty moving out of poverty.

As the lot of the disadvantaged improves, the lot of the advantaged improves, much like, *though the opposite of,* trickle-down economics fostered by some. Trickle-down economics is supply side economics whereas trickle-up is demand side economics. With trickle-down economics the benefits go first to the advantaged and then the disadvantaged. This approach is used now and has been used during much of the past. With trickle-up economics the benefits go to the disadvantaged and then the advantaged; benefits move up the chain, much like transpiration (the process of water traveling from the roots of a tree up through the trunk, the limbs, and then to the leaves). Nature has proven this works. It is worth trying.

The American dream needs to be reinvigorated. The approach described here is in answer to Senator Carl Levin's concern who said in his farewell address on the floor of the US Senate on December 17, 2014, that the "growing gulf between the fortunate few and the struggling many is a threat to the dream that has animated this nation since its founding". A similar comment may be made regarding the peoples of every country in the world. The realization of that dream becomes more likely, today, with the realizable objectives indicated in this book.

The Chinese say that one can destroy a person with a single painful event or many small injuries. The opposite is also true; one can build a person with kindness and warmth or with a thousand praises. Parents can encourage their children with praise and warmth or demean them with the many forms of neglect. Society can, similarly, do either. Depending on what they hear, children may conclude, "I am black. I am stupid." Or, "I am okay. I can achieve." It may even suggest to girls, "Of course I can do math." Society and parents have responsibilities to honor diversity, not denigrate differences. Individuals and parents have the responsibility to perceive and convey the positive.

Parents need to set the path; where possible, to create opportunities. They cannot simply blame the system or society or the schools for acknowledged weaknesses. Where appropriate, they surely must do that, but they must also look within. Parents and society need to help children reach their potential; provide environmental opportunities for their children to more closely reach their potential. Paraphrasing Hillel, if they don't do it, if they do not take the lead, who will? If not now, then when[141]?"

Eleanor Roosevelt, who much of her life struggled with feelings of low self-esteem, said, "No one can make you feel inferior without your consent". Expanding on this, Rabbi Louis L. Mann said, "What happens to a man is less significant than what happens within him." It is not enough to enhance intelligence, per se, or to be aware that one can enhance intelligence. One must also address motivation, interest, self-confidence, and hope. Joan Didion, American novelist and journalist wrote on the same theme, "The willingness to accept responsibility for one's life is the source from which self-respect springs." But that is not always easy, as the German philosopher Heine observed, "When an individual endeavors to lift himself above his fellows, he is dragged down by the mass, either by means of ridicule or by

---

[141] Ethics of the Fathers

calumny." The challenges to reach beyond one's supposed station are great, but the rewards are beyond expectations.

Not so strangely, these considerations apply to the disadvantaged as well as the advantaged. Many, of all stripes, feel limited hope, and accordingly, not surprisingly, become frustrated and indifferent bystanders to their own life. To be an active self-advocate requires optimism and the sense of real potential. For people to advance themselves has always been difficult and often fruitless. However, knowing that opportunities open-up with greater capability, and that capability can be improved with enhanced intelligence, and that intelligence can be substantially improved, especially among the disadvantaged, striving to reach closer to one's potential should be much easier and should be much more likely.

The process presents an upward spiral; as more people enter the middle-class the demand for goods and services increases, causing more jobs, and more people to enter the middle class, which results in yet more demand for goods and services; the upward spiral develops momentum. In parallel, as the middle-class grows, more benefits are realized by those providing goods and services, more of the middle-class protects their own health and wealth and that of society, more profits are realized in providing the fruits of society, more of the middle-class protects their own wealth and health and the health of society, and so, worth repeating, the spiral finds its own momentum across the entire society.

The rather undesirable scenarios painted by Charles Murray and others, of gated communities locking-out the poor and locking-in the rich, need never occur. Though poverty may never disappear, it can be dramatically reduced, largely eliminated. Diversity in cultures and in nationalities will become more welcome and more respected as the planet becomes smaller in size and larger in possibilities.

Poverty, as we know it today, in the numbers we know today, need not be here to stay.

A severe and subtle tyranny of time often takes hold. The advantaged, being comfortable, have little need to make changes; time is in their favor. In contrast, time is the villain of the disadvantaged; the longer it takes to improve their lot, the longer injustices plague their lives. The tyranny of time often favors the status quo; it favors the already rich, while the inertia of change prevents the poor from moving forward. There are powerful reasons to change the status quo now. To quote Martin Luther King who said in 1963, over a half century ago, there is a fierce urgency of now.

It is suggested that unlike other issues being considered by society, the potential described in this book presents huge benefits for both the disadvantaged and advantaged in the Western world and the world over; it benefits the advantaged, including the richest of the advantaged with financial benefits, reduced societal problems, and increased national and international tranquility. If the advantaged were to take the time to understand the vast implications to them, they would be disturbed to find that huge benefits, not merely profits, go unrealized every month that the suggested interventions are not implemented.

Convincing the advantaged to actively foster upward mobility for the billions of people seeking to take part in the fruits of society is vital. Interestingly, many of the wealthy, such as the Gates family and the Buffet family, are essentially in general agreement with these notions; Bill Gates said the most important part of his work is to convey that every life has the same value. But then there are others with extreme wealth that still choose to have nothing to do with this kind of thinking. They believe that independence and drive is all that one needs to achieve success. And then there are the many unaware or indifferent to the possible improvement of the disadvantaged. On a more optimistic note, anthropologist Jan Goodal declared that informed people care, and those that care, act. This book was written to inform people of what is reasonable, hoping that consequently they would act accordingly.

Some will look at these observations and respond that many of the poor are not willing to apply the effort to get out of poverty. Surely, some take handouts and choose not to work, but if the data were closely examined, this number represents a tiny percentage of the whole, perhaps as large as the percentage of the rich who also live on handouts though of a dramatically different kind and magnitude. Living on the dole is not a monopoly of the disadvantaged; some of the rich have developed the ability to an art form. Today the rich benefit from vast governmental subsidies. The wealthy have the power to lobby for these financial benefits. While both these types of handouts need review and modification, this book offers a way to reduce both poverty and the need for continued but gradually reduced welfare. While this author does not care that the rich get richer, the clear focus is that the disadvantaged achieve closer to their potential; that the poor move out of poverty into the middle-class.

There will always be relatively poor people, but there need not be so many, and, they need not be as poor. There will always be less capable people, but there need not be so many, and there need not be so much disparity between the haves and the have-nots. Throughout this book I have included several real stories of children burdened by poverty, by disadvantages, or by social baggage. I conveyed these accounts because, it seems, people are touched by the pain of one person; they have less sensitivity, understanding, for the pain of thousands of people, and even much less when we talk of the poverty of billions of people, which is what we are talking about in this book. The greater the magnitude of a problem, the more indifferent people seem to be. On the other hand, each of these billion children needs and feels love, has a tender smile, a unique history, and has a similar potential – a future over which we, each one of us, has influence, if we choose to exercise it.

Every child mentioned in this book has real feelings and gives genuine affection; many of them hugged me when I saw them in school, and years later would run across the street to greet me. Dottie, who was wrongly placed in resource classes, graduated from high school and then college. One child called me years later and asked for help to get her GED, which she did and went on to nursing school. One child called me from the hospital so that I attend the birthing of her first child. To my happy surprise, Bill, who wanted to drop out of school as a sophomore in high school, came unexpectedly to my home just to say hi; he had become a firefighter, and was considering starting his own business. One child invited me to his wedding. Jack, whose teacher said, "You do not understand – he cannot," graduated from high school. Bob, who was the school tough, went on to parochial school, and probably did very well. Truth be told, though I am glad to have helped many children, I did not help all the children I encountered, nor, of course, the many more that I didn't.

Research studies are usually documented, peer-reviewed, published, and then, all-too-often lost in the literature. This book has sought out appropriate studies in the literature to support a solution of a wide spectrum of societal problems. This book has shown that a single solution—improving the intelligence, most substantially of the disadvantaged—can dramatically reduce a myriad of societal problems. Some of the interventions suggested to enhance intelligence are partly in place but need to be improved by, for example, providing not only free or price-reduced lunch and breakfast while school is in session but also when it's not, and also providing mineral and vitamin supplements to pregnant women during their offspring's prenatal stage but also the developmental stages to follow. Others such as parenting training and very early and quality education need to find acceptance in the United States.

With widely acknowledged evidence, we now can unequivocally say that intelligence may be significantly enhanced, most substantially of the disadvantaged; the IQs of those toward the low end of the IQ spectrum may be raised to the near-average range and higher. Additionally, most of these objectives are essentially self-supportive; the costs are more than covered by the immediate savings in related areas. Furthermore, with enhanced intelligence comes greater capability and thus increased opportunity, and accordingly, it is reasonable to conclude that poverty may be substantially reduced – perhaps even largely eradicated.

The purpose of this book is to encourage a conversation—an honest, open-minded dialogue—to discuss the merits and weaknesses of interventions that increase intelligence and thereby decrease inequalities in opportunities, as well as decrease economic inequalities.

In parallel, it is suggested that we establish an extensive research activity, much like the Genome Project, to better understand intelligence. Such an activity would likely also achieve agreement on myths to be moved into the field's mythology archives so that advances in understanding intelligence may be more effectively made. The interventions suggested in this book need not wait for resolutions of these academic issues because we currently know that these interventions work.

While it is not possible to achieve equality in opportunity or economic equality, inequality can be minimized; great benefits are realized as poverty is minimized and as people live in reasonable comfort.

However, to achieve the full fruition of these objectives, essentially all need to participate. Indifference or complacency of any group limits success. The advantaged, including the very rich, must see the apparent benefits and play a part in achieving the acknowledged objectives. Of course, those in poverty must strive to achieve as well. A powerful cycle takes hold. Societies throughout the world should see dramatic benefits. Stability would significantly increase while conflict would substantially decrease. Many more will have reason to be optimistic for themselves and their children.

Hopefully, any faults in the envisioned tapestry may be recognized and corrected as the tapestry is developed. With rational discussion, the actual tapestry may result in coherence, as a work of beauty, simplicity, and endurance.

For thousands of years, we have had reason to believe that poverty was part of societal structure; essentially nothing could be done to prevent poverty. This view is no longer valid. While, certainly, there are many causes of poverty, *the single most important and widespread cause of poverty is the environmental constraints on intelligence*; inadequate nutrition, just one such environmental constraint typical of the poor, limits the normal development of intelligence, constrains effective education, thus limits employability, and accordingly causes poverty. A circular pattern has evolved and stubbornly continues to exist, namely,

Constrained intelligence causes poverty,
and
poverty constrains intelligence.

Today, we know how to break this historically persistent circular pattern. As shown in this book, we know many interventions that decrease environmental constraints on intelligence in all developmental stages of life, thereby enhancing capability and intelligence, especially of the disadvantaged. Reducing these constraints on intelligence, thus enhances capability, increases employability, and permits more people to move into the middle-class, to participate with and enjoy the fruits of society, to seek more goods and services, resulting in even more people needed in the middle-class, and the upward spiral gains momentum.

Additionally, there is a strong relationship between intelligence and school dropout rate, employability, criminality, and between intelligence and many other societal problems. Not surprisingly, as intelligence increases societal problems decrease as do their substantial costs. Reducing the constraints on intelligence, thereby enhancing intelligence, especially of the poor, dramatically reduces poverty, and many other societal issues.

Intelligence, probably the most distinguishing human characteristic, is largely wasted among most young minds in not only the developing world but surprisingly in the developed world as well; and as young minds go, so go adult minds. But intelligence and capability of most humanity need not continue to be wasted. Most people across the planet can far more closely reach their potential intelligence. Enhancing intelligence of the disadvantaged not only benefits the poor but as the middle class grows, demand for goods and services increases benefiting, of course, those providing goods and services.

Some will feel no urgency, but others will cry out, if not now, then when? The status quo works fine for the advantaged; change can wait. However, for the disadvantaged the current condition is painful and hopeless – time works against the disadvantaged. For the benefit of all, and especially those in poverty, *now* is the only hopeful, reasonable, and ethical answer. *Soon* is not soon enough. No part of anyone's life should ever be deliberately wasted.

The challenge in writing this book was convincing the poor that the objectives are within their reach, convincing the rich that this was in their best interests, and convincing policy makers to make it happen.

References

Abel, E. L. (1998). Prevention of Alcohol Abuse-Related Birth Effects Public Education Efforts. *Alcohol & Alcoholism* Vol. 33, No. 4, pp. 411—416,

Abel, E. L., Hannigan, J. H. (1995). Maternal Risk Factors in Fetal Alcohol Syndrome: Provocative and Permissive influences. *Neurotoxicol Teratol.* Jul-Aug; 17(4): 445-62.

Achenbach, T. M., Phares, V., Howell, C. T., Rauh, V. A., and Nurcombe, B. (1990). Seven-Year Outcome of the Vermont Intervention Program for Low-Birthweight Infants. *Child Development.* 61, 1672-1681

Adler, I. (2014). How Childhood Neglect Harms The Brain. Retrieved from url http://commonhealth.wbur.org/2014/06/trauma-abuse-brain-matters

Almond, D., Chay, K. Y., Lee, D. S. (2005). The Costs of Low Birth Weight. *The Quarterly Journal of Economics* 120 (3): 1031-1083. doi: 10.1093/qje/120.3.1031

Atkinson, A. (2015). *Inequality: What can be done.* Harvard University Press

Barnett, W. S. (1995). Long-Term Effects of Early Childhood Programs on Cognitive and School Outcomes. *The Future of Children,* 5(3), 25–50. Retrieved from http://www.jstor.org/stable/1602366

Bartick, M., and Reinhold, A. (2010). "The Burden of Suboptimal Breastfeeding in the United States: A Pediatric Cost Analysis." *Pediatrics*

125, no. 5 (April 5, 2010): e1048–e1056. http://pediatrics.aappublications.org/cgi/doi/10.1542/peds.2009-1616 .

Bayless, T. M. (1995). Milk intolerance: Clinical development and epidemiological aspects. In I. I. Gottesman and L.L. Heston (Eds.) *Summary of the Conference on Lactose and Milk Intolerance.* DHEW Publication No. (OCD) 73-19. Washington, D.C.: U.S. Department of Health, Education, and Welfare, 1973. Pp.10-18.1995

Benton, D. (2001). Micro-nutrient supplementation and the intelligence of children. *NeuroScience and Biobehavioral Reviews,* 25, 297-309

Block, N. (1995). How Heritability Misleads About Race. *The Journal of Cognition.* Bloom 1964 p.215 source Erickson, E. H., 1950. Childhood and society. New York: Norton

Bloom, B. S. (1964). *Stability and Change in Human Characteristics.* New York; Wiley.

Boas, F., (1911). Changes in bodily form of descendants of immigrants. Washington: U.S. Senate Document, No. 208

Bodmer, W. F. (1999). Race and IQ: The genetic background. In *Race and IQ* (Expanded Edition) ed. A. Montagu. 1999. Oxford Univ. Press. Oxford, G.B. pp. 308-342.

Bouchard, T. J., Jr., & McGue, M. 1981. Familial studies of intelligence: A review. *Science,* 212. 1055-59.

Brinch, C. N. & Galloway, T. A. (2012). Schooling in Adolescence raises IQ scores. PNAS. vol. 109 no. 2, 425-430 doi: 10.1073/pnas.1106077109

Bronfenbrenner, U. (1974). Nature and Nurture: A reinterpretation of the evidence republished. In *Race and IQ* (ed.) by Ashly Montagu. 2002 Oxford Univ. Press. Oxford

Bronfenbrenner, U. (1974). Is Early Intervention Effective? Some Studies of Early Education in Familial and Extra-Familial Settings. In *Race and IQ* (ed.) by Ashly Montagu. 2002 Oxford Univ. Press. Oxford

Bronfenbrenner, U. (2002). Is early intervention effective? Some studies of early education in familial and extra-familial settings. In A. Montagu (Ed.), *Race and IQ*. Oxford, UK: Oxford University Press.

Burt, C., Jones, E., Miller, E., & Moodie, W. (1934). *How the mind works.* New York: Appleton-Century-Crofts

Campbell, F. A; Conti, G. Heckman, J. J.; Moon, S.: Pungello, E. P. 2014. Early Childhood Investments Substantially Boost Adult Health, *Science*, (343): pp. 1478-1485.

Cauthen, N. K. and Fass, S. (2008). Measuring Poverty in the United States

Ceci, S. J. and Liker, J. (1986). "A Day at the Races: A Study of IQ, Expertise, and Cognitive Complexity." *Journal of Experimental Psychology*: General, 115, 255-266.

Ceci, S. J. and Liker, J. (1986). "Academic and Non-Academic Intelligence: An Experimental Separation." In R. J. Sternberg and R. K. Wagner (eds.), *Practical Intelligence: Origins of Competence in Everyday World.* New York: Cambridge Univ. Press.

Ceci, S. J. and Liker, J. 1988. "Stalking the IQ-Expertise Relationship: When the Critics Go Fishing." *Journal of Experimental Psychology: General*, 117, 96-100.

Ceci, S. J. (1991). How much does schooling influence general intelligence and its cognitive components? A reassessment of the evidence. *Developmental Psychology, 27*, 702–722. doi:10.1037/0012-1649.27.5.70

Centers for Disease Control, National Center for Health Statistics, National Vital Statistics System. (n.d.). *Gestation and Birthweight - 2009.* Retrieved

from http://www.cdc.gov/nchs/data_access/vitalstats/VitalStats_
Births.htm

Chambers, B., & Paynter, S. (2012, January 16). Poverty in N.C.: The real
numbers. *The News and Observer*, p. 9A.

Chomsky, N. 1975. *Reflections on language*. New York: Pantheon

Colom, R., Lluis-Font, J. M., Andre's-Pueyo, A. (2005). The generational
intelligence gains are caused by decreasing variance in the lower half
of the distribution: Supporting evidence for the nutrition hypothesis.
Intelligence 33 (2005) 83–91

Cowen, T. (2014). Income Inequality Is Not Rising Globally. It's Falling. NY
Times Economic View, July 19, 2014

Currie, J. (2001). Early Childhood Education Programs. *The Journal of
Economic Perspectives*, 15(2), 213–238. Retrieved from http://www.jstor.
org/stable/2696599

Currie, J. (2007). How Should We Interpret the Evidence about Head Start?
Journal of Policy Analysis and Management, 26(3), 681–684. Retrieved
from http://www.jstor.org/stable/30163424

Currie, J., & Thomas, D. (1995). Does Head Start Make a Difference? *The
American Economic Review*, 85(3), 341–364. Retrieved from http://
www.jstor.org/stable/2118178

Currie, J., & Thomas, D. (1998). School Quality and the Longer-Term Effects
of Head Start. *National Bureau of Economic Research Working Paper
Series, No. 6362*. Retrieved from http://www.nber.org/papers/w6362

Darwin, C. (1839). *Voyage of the Beagle*. J. M. Dent & Sons, Ltd. N.Y., N.Y.,
E. P Dutton & Co., Inc. (1936) xvi, 496 p.

Degroot, A. D. (1951). War and the intelligence of youth. *Journal of Abnormal and Social Psychology, 46,* 596–597. doi:10.1037/h0057014

DeNavas-Walt, C., Proctor, B. D., Smith, J. C., & U.S. Census Bureau. (2010). *Income, Poverty, and Health Insurance Coverage in the United States: 2009* (No. P60-238). Current Population Reports. Washington, DC: U.S. Government Printing Office. Retrieved from http://www.census.gov/prod/2010pubs/p60-238.pdf

Dickens, W. T. & Flynn, J. R. (2006a). Black Americans reduce the racial IQ gap: Evidence from standardization samples. *Psychological Science 16* (10): 913–920. doi:10.1111/j.1467-9280.2006.01802.x

Dickens, W. T., & Flynn, J. R. (2006b). Common Ground and Differences. *Psychological Science, 17*(10), 923–924. doi:10.1111/j.1467-9280.2006.01804.x

Dillon, S. (2009, October 9). Study Finds High Rate of Imprisonment Among Dropouts . The NY Times. Retrieved from url

Dweck, C. S. (2006). *Mindset: the new psychology of success.* Random House, N. Y., N. Y.

Eckberg, D. L., (1979). *Intelligence and race: The origins and dimensions of the IQ controversy.* Praeger Publishers, NY, NY.

Economist June 1st, 2013, *Towards the end of poverty; The world's next great leap forward.* (2013, June 1). *The Economist, 407*(8838), 11(US). Retrieved from http://go.galegroup.com/ps/i.do?id=GALE%7CA331787861&v=2.1&u=duke_perkins&it=r&p=AONE&sw=w&asid=588bfd967a6cedd9af601402c5db557b

Economist, June 6th, 2015, *Anthony Atkinson, the godfather of inequality research, on a growing problem.*

Eppig, C., (2010). *Parasite prevalence and the worldwide distribution of cognitive ability. Proceedings of the Royal Society B* doi:10.1098/rspb.2010.0973

Eyferth, K. (1961). Leistungern verschiedener Gruppen von Besatzungskidern in Hamburg-Wechsler Intelligenztest fur Kinder (HAWIK). *Archiv Fur die gesamte Psychologie,* 113, 222-241.

Eysenck. H. J. (1991). Race and intelligence: An alternative hypothesis. *Mankind Quarterly,* 32, 123–125.

Fisher, G. M. (1992a.) Poverty Guidelines for 1992. *Social Security Bulletin,* 55(1), 43–46. Retrieved February 24, 2011 from http://aspe.hhs.gov/poverty/papers/background-paper92.shtml

Fisher, G. M. (1992b.) The development and history of the poverty thresholds. *Social Security Bulletin,* 55(4), 3–16. Retrieved February 24, 2011 from http://www.ssa.gov/history/fisheronpoverty.html

Flynn, J. R. (1980). *Race, IQ, and Jensen.* London: Routledge & Kegan Paul.

Flynn, J. R. (1984). The mean IQ of Americans: Massive gains 1932 to 1978. Psychological Bulletin, 95, 29-51

Flynn, J. R. (2006). TETHERING THE ELEPHANT; Capital Cases, IQ, and the Flynn Effect. *Psychology, Public Policy, and Law* 2006, Vol. 12, No. 2, 170–189

Flynn, J. R. (2008). *Where Have All the Liberals Gone? Race, Class, and Ideals in America.* Cambridge Univ. Press.

Fodor, J. A. 1983. *The modularity of mind.* Cambridge, Mass.: MIT Press.

Garber, H. L. (1988). *The Milwaukee Project: Preventing Mental Retardation in Children at Risk.* Washington, D. C.: American Association on Mental Retardation

Gardner, H. 1987. "Developing the Spectrum of Human Intelligence." *Harvard Educational Journal*, 57, 187-193.

Gertler, P., Heckman, J., Pinto, R., Zanolini, A., Vermeersch, C., Walker, S., Chang, S. M., Grantham-McGregor, S. (2014). Labor market returns to an early childhood stimulation intervention in Jamaica. *Science*, Vol. 344, no. 6187, pp. 998-1001, DOI: 10.1126/science.1251178

Gould, S. J. (1981). *The Mismeasure of Man*. W. W. Norton: New York, N.Y

Härnquist, K. (1968). Relative changes in intelligence from 13 to 18. *Scandinavian Journal of Psychology*, 9, 50–64. doi:10.1111/j.1467-9450.1968.tb00519.x

Harrell, R. F., Woodyard, E., & Gates, A. J. (1955). *The effect of mothers' diets on the intelligence of offspring: A study of the influence of vitamin supplementation of the diets of pregnant and lactating women on the intelligence of their children*. Bureau of Publications. Teachers College, Columbia University. N. Y., N. Y.

Harris, J. R. (1998). *The Nurture Assumption: Why children turn out the way they do*. Touchstone Book. New York, NY.

Hawking, S. (1998). *A Brief History of Time; Updated and Expanded Tenth Anniversary Edition*. Bantam Books, New York, NY.

Hayes, D. P., and Grether, J. (1982). The school year and vacations: When do students learn? Cornell *Journal of Social Relations*, 17, 56–71.

Heber, R. (1969). *Rehabilitation of Families At Risk for Mental Retardation*. Regional Rehabilitation Center, University of Wisconsin

Herrnstein, R. J., and Murray, C. (1994). *The Bell Curve: Intelligence and the class structure in American life*. Simon and Schuster. New York, NY.

Herrnstein, R. J., Nickerson, R. S., DeSanche, M., and Swets, J. A. (1986). Teaching thinking skills. *American Psychologist, 41,* 1279–1289. doi:10.1037/0003-066X.41.11.1279

Heyns, B. L. (1978). *Summer learning and the effects of schooling.* New York: Academic Press.

Horta, B. L., Loret de Mola, C., Victora, C. G. (2015). Breastfeeding and intelligence: a systematic review and meta-analysis. *Acta Paediatr.* 2015 Dec;104(467):14-9. doi: 10.1111/apa.13139.

Husen, T. (1951). The influence of schooling upon IQ. *Theoria, 17,* 61–88. doi:10.1111/j.1755-2567.1951.tb00233.x

Hunt, J. McV. (1961). Intelligence and experience. The Ronald Press Co. NY, NY

Jacobs, A. July 8, 2018. U.S. Opposition to Breast-Feeding Resolution Stuns World Health Officials. NY Times.

Jacobson, J. L. and Jacobson, S. W. (1992). Breast feeding and intelligence. *Lancet,* 339, 926

Jacobson, J. L. and Jacobson, S. W. (2006). Intellectual Impairment in Children Exposed to Polychlorinated Biphenyls in Utero. *N Engl J Med* 1996; 35:783-789 September 12,1996DOI: 10.1056/NEJM199609123351104

Jacoby, R., Glauberman, N., (1995). *The Bell Curve Debate, History, Documents, Opinions.* Times Books. N.Y., N.Y. p.126, 7

Jencks, C., & Phillips, M. (Eds). (1998) *The Black-White test score gap.* Washington, DC: Brookings Institute Press.

Jencks, C., Smith, M., Aclad, H., Bane, M. J., Cohen, D., Gintis, H., Heynes, B., and Mitchelson, S. (1972). *Inequality: A reassessment of the effect of family and schooling in America*. New York: Basic Books.

Jensen, A. R. (1969). How much can we boost IQ and scholastic achievement? *Harvard Educational Review*. 39:1–123.

Jensen, A. R. (1972). Genetics and education. Harper & Row, N.Y., N.Y.

Jensen, A. R. 1973. 'Educability and Group Differences'. New York. Harper & Row. P. 363

Jensen, A. R. (1998). *The g-factor: The science of mental ability*. Praeger. Westport, Conn.

Johnson, S. (2001, Dec.) *Cost of infant feeding*. San Diego Breastfeeding Coalition. http://www.breastfeeding.org/bfacts/cost.html

Kalil, A. & Ryan, (2010). Mothers' Economic Conditions and Sources of Support in Fragile Families. *The Future of Children*. ISSN 1054-8289.

Karnes, M. B., Hodgins, A., and Teska, J. A. (1968). An evaluation of two preschool programs for disadvantaged children: A traditional and a highly structured experimental preschool. *Exceptional Children, 34,* 667–676.

Karnes, M. B., and Badger, E. E. (1969.) Training mothers to instruct their infants at home. In M. B. Karnes, *Research and development program on preschool disadvantaged children: Final report*. Washington, D. C.: U. S. Office of Education.

Klein, N. (2014). *This Changes Everything; Capitalism vs The Climate*. Simon and Schuster, NY. NY.

Kristoff, N., WuDunn,. S. (2014). *The Way To Beat Poverty*. NY Times Sunday Review, September 12, 2014,

Kristoff, N. (2014, November 19). Do Politicians Love Kids? Retrieved from www.New York Times.comLazar, I., Darlington, R., Murray, H., Royce, J., Snipper, A., & Ramey, C. T. (1982). Lasting Effects of Early Education: A Report from the Consortium for Longitudinal Studies. *Monographs of the Society for Research in Child Development*, 47(2/3), i–151. Retrieved from http://www.jstor.org/stable/1165938

Lewontin, R. C., Rose, S., and Kamin, L. J. (1984). *Not in our genes.* Pantheon Books: NY, NY

Liker, J., and Ceci, S. J. 1987. "Role of Experience in IQ." *Journal of Experimental Psychology*, 116, 304-306.

Loehlin, J. C., Lindzey, G., & Spuhler, J. N. (1975). *Race differences in intelligence.* W. H. Freeman and Co. San Francisco, Calif.

Lynn, R. (1998). In support of the nutrition theory. In U. Neisser (Ed.), *The Rising Curve: Long-Term Gains in IQ and Related Measures* (pp. 207–218). American Psychological Association. Washington, DC.

Lynn, R., and Vanhanen, T. (2002). IQ and the wealth of nations. Westport, CT: Praeger.

Lynn, R. and Vanhanen, T. (2006). *IQ and global inequality.* Washington Summit Publishers, Augusta, Ga.

Mackey, A. P., Finn, A. S., Leonard, J. A., Jacoby-Senghor, D. S., West, M. R., Gabrieli, C. F. O., Gabrieli, J. D. E., (2015). "Neuroanatomical Correlates of the Income-Achievement Gap", Psychological Science, April 20, 2015, 0956797615572233

Magnuson, K. A. & Waldfogel, J. (2005). Early childhood care and education: effects on ethnic and racial gaps in school readiness. *The Future of Children*, 15(1), 169+. Retrieved from http://go.galegroup.com/ps/i. do?id=GALE%7CA136653456&v=2.1&u=duke_perkins&it=r&p= AONE&sw=w&asid=d0c2a88b3f4b690f06c45493e2d48109

Marks, D. F. (2010). IQ Variations Across Time, Race, and Nationality: An artifact of differences in literacy skills. *Psychological Reports* Volume 106, Issue 3, 643–664.

Martorell, R. (1998) Nutrition and the worldwide rise in IQ scores. In U. Neisser (Ed.), *The Rising Curve: Long-Term Gains in IQ and Related Measures* (pp. 183–206). American Psychological Association. Washington, DC.

McDowell, M. A., Wang, C., & Kennedy-Stephenson, J. (2008). *Breastfeeding in the United States: Findings from the National Health and Nutrition Examination Surveys 1999-2006* (CHS Data Briefs No. 5). Hyattsville, MD: National Center for Health Statistics. Retrieved from http://www.cdc.gov/nchs/data/databriefs/db05.htm

McGillicuddy-DeLisi, A. V. (1982). The relationship between parents' beliefs about development and family constellation, SES, and parents' teaching strategies. In L. M. Laosa & I. E. Sigel (Eds.), *Families as learning environments for children* (pp. 1–45). New York: Plenum Press.

Moore, E. G. J. (1986). Family socialization and the IQ test performance of traditionally and transracially adopted Black children. *Developmental Psychology, 22*, 317–326. doi:10.1037/0012-1649.22.3.317

Murray, C. (2012). *Coming Apart: The State of White America, 1960-2010.* Random House, Inc., New York, NY.

Muschkin, C. G., Ladd, H. F., and Dodge, K. A. (2015). Impact of North Carolina's Early Childhood Initiatives on Special Education Placements in Third Grade. *American Educational Research Association's journal, Educational Evaluation and Policy Analysis,* Month 201X, Vol. XX, No. X, pp. 1–23

National Scientific Council on the Developing Child. (2012). *The Science of Neglect: The Persistent Absence of Responsive Care Disrupts the Developing Brain: Working Paper 12.* http://www.developingchild.harvard.edu

Neisser, U., Boodoo, G., Bouchard, T. J. J., Boykin, A. W., Brody, N., Ceci, S. J., ... Urbina, S. (1996). Intelligence: Knowns and unknowns. *American Psychologist*, 51(2), 77–101. doi:10.1037/0003-066X.51.2.77

Nelson, C. A., Fox, N. A., and Zeanah, C. H. (2014). Romania's Abandoned Children: Deprivation, Brain Development and the Struggle for Recovery, Harvard University Press, 416pp, ISBN 9780674724709 and 26079 (e-book)

Newsweek 9/19/2005, p. 59 SOCIETY: "Boy Brains, Girl Brains" (p. 59). General Editor Peg Tyre. http://www.msnbc.msn.com/id/9285515/site/newsweek/

Nisbett, R. E. (2005). Heredity, environment, and race differences in IQ: A commentary on Rushton and Jensen. *Psychology, Public Policy and Law*, 11, 302–310. doi:10.1037/1076-8971.11.2.295

Nisbett, R. E. (2009). *Intelligence and how we get it; Why schools and culture count*. W. W. Norton & Company, New York, NY.

Noble, K. G., Houston, S. M., Brito, N. H., Batsch, H., Kan, E., Kuperman, J. M., Akshoomoff, N., Amaral, D. G., Bloss, C. S., Libiger, O., Shork, N. J., Murray, S. S., Casey, B. J., Chang, L., Earnst, T. M., Frazier, J. A., Gruen, J. R., Kennedy, D. N., Zijl. P. V., Mostofsky, S., Kaufmann, W. E., Kenet, T., Dale, A. M., Jernigan, T. L., Sowell, E. R. (2015). Family income, parental education and brain structure in children and adolescents. *Nature Neuroscience*. Vol.18, Num.5 doi: 10.1038/nn 3983

Ortiz-Mantilla, S., Choudhury, N., Leevers, H., & Benasich, A. A. (2008). Understanding language and cognitive deficits in very low birth weight children. Article first published online: 19 FEB 2008. DOI: 10.1002/dev.20278

State of North Carolina, Department of Health and Human Services, North Carolina State Center for Health Statistics. (n.d.). *2009 North Carolina resident live births* (Annual report first published for 2001.).

Raleigh, NC. Retrieved from http://www.schs.state.nc.us/schs/births/matched/2009/

Piketty, T. (2014). *Capital in the Twenty-First Century,* The Belknap Press of Harvard University Press, Cambridge, Mass.

Pinker, S. (2002) *The Blank Slate: The Modern Denial of Human Nature.* Viking. New York, NY.

Robert D. Putnam, R. D. (2015). *Our kids: The American Dream in Crises.* Simon & Schuster, N.Y., N.Y.

Quinn, P., O'Callaghan, M., Williams, G., Najman, J., Andersen, M., & Bor, W. (2001). The effect of breastfeeding on child development at 5 years: A cohort study. *Journal of Pediatrics and Child Health, 37*(5), 465-469. doi:10.1046/j.1440-1754.2001.00702.x

Ramey, S. L., and Ramey, C. T. (1999) Early experience and early intervention for children "at risk" for developmental delay and mental retardation. *Mental Retardation and Developmental Disabilities Research Reviews, 5,* 1–10.

Reed, E.W. and Reed, S.C. 1965 *Mental retardation: a family study.* Philadelphia, PA: Saunders.

Reich, R. (Fri Nov 07, 2014 at 08:47 AM PST). *Why Dems Lost The Election And What They Can Do About It.* byGleninCAFollow forGleninCA

Riordan, J. M. "The Cost of Not Breastfeeding: A Commentary." *Journal of Human Lactation* 13, no. 2 (June 1, 1997): 93–97. http://jhl.sagepub.com/cgi/doi/10.1177/089033449701300202

De Rivero, Oswaldo. (2001). *The Myth of Development; Non-Viable Economies and the Crises of Civilization,* Zed Books, London ISBN 1 85649 949 9 Pb,

Rollins, N. C., Bhandari, N., Hajeebhoy, N., Horton, S., Lutter, C. K., Martines, J. C., Piwoz, E. G.,

Richter, L. M., Victora, C. G., on behalf of The Lancet Breastfeeding Series Group. (2016). Why invest, and what it will take to improve breastfeeding practices? *Lancet.* Volume 387, Issue 10017, 30 January–5 February 2016, Pages 491-504

Rushton, J. P. (1995). *Race, Evolution, Behavior.* New Brunswick, NJ: Transaction Publishers.

Rushton, J. P., & Jensen, A. R. (2005), Thirty years of research on race differences in cognitive ability, *Psychology, Public Policy, and Law* **11**: 235–294, 406. doi: 10.1037/1076-8971.11.2.235

Rushton, J. P., & Jensen, A. R. (2006). The Totality of Available Evidence Shows the Race IQ Gap Still Remains. *Psychological Science, 17*(10), 921–922. doi:10.1111/j.1467-9280.2006.01803.x

Rushton, J. P., & Jensen, A. R. (2010). Race and IQ: A theory-based review of the research in Richard Nisbett's *Intelligence and how to get it. The Open Psychology Journal, 3,* 9–35. http://www.benthamscience.com/open/topsyj/articles/V003/9TOPSYJ.htm

Sanders, B. K., (1934). *Environment and Growth.* Baltimore: Warwick and York.

San Diego Breastfeeding Coalition, Prepared by Sharon Johnson, RN, IBCLC Dec. 2001.

Scarr, S. (1981). *Race, social class, and individual differences in IQ.* Lawrence Erlbaum Assoc. Hillsdale, NJ.

Scarr, S. 1998. How do families Affect Intelligence? Social Environmental and Behavior Genetic Predictions. *Human Cognitive Abilities in Theory and Practice* 1998 p125

Scarr, S., & Weinberg, R. A. (1976). IQ test performance of black children adopted by white families. *American Psychologist, 31*, 726-739.

Shockley, W. (1971). Morals, mathematics, and the moral obligation to diagnose the origin of Negro IQ deficits. *Review of Educational Research, 41*, 369–377.

Shockley, W. (1972). Dysgenics, geneticity, raciology: A challenge to the intellectual responsibility of educators. *Phi Delta Kappan, 53*, 297–307.

Shuey, A. M. (1966). *The testing of negro intelligence.* New York, NY.

Sigman, M. and Whaley, S. E. (1998). The Role of Nutrition in the Development of Intelligence. In U. Neisser (Ed.), *The Rising Curve: Long-Term Gains in IQ and Related Measures* (pp. 155–182). American Psychological Association. Washington, DC.

Snyderman, M. & Rothman, S. (1987). Survey of Expert Opinion on Intelligence and Aptitude Testing. *American Psychologist.* Vol. 42, No. 2, 137-144.

Snyderman, M., & Rothman, S. (1988). *The IQ Controversy, the Media and Public Policy.* Transaction Books. New Brunswick, NJ.

Spitz, R. A., Wolf, K. M. (1946). *Grief: A Peril in Infancy.* N. Y. University Film Library. N.Y., N.Y.

Sternberg, R. J., Powell, C., McGrane, P., & Grantham-McGregor, S. (1997). Effects of a Parasitic Infection on Cognitive Functioning. *Journal of Experimental Psychology: Applied.* Vol. 3, No. 1, 67-76.

Stiglitz. J. E. (2014). White Paper; *Reforming Taxation to Promote Growth and Equity*

Sullivan, K. M., Ford, E. S., Azrak, M. F., Mokdad, A. H. (2009). Multivitamin use in pregnant and nonpregnant women: results from the Behavioral Risk Factor Surveillance System. *Public Health Rep* 124: 384–90.

Sulloway, F. J. (2007). Perspective PSYCHOLOGY: Birth Order and Intelligence, *Science 317*, 1711-1712.

Sum, A., Khatiwada, I., McLaughlin, J., Palma, S. (2009). The consequences of dropping out of high school joblessness and jailing for high school dropouts and the high cost for taxpayers. *Center for Labor Market Studies Publications*. Paper 23. http://hdl.handle.net/2047/d20000596

Thompson, W. R., & Heron, W. 1954. The effects of restricting early experiments on the problem-solving capacity of dogs. *Canadian. J. Psychol.* [8, 17-31, 103-106, 139, 155, 263-265, 269, 316, 353, 356].

Turkheimer, E. (2000). Three Laws of Behavioral Genetics and What They Mean. *American Psychological Society. Vol. 9*, Num. 5.

Turkheimer, E., Haley, A., Waldron, M., D'Onofrio, B., & Gottesman, I. I. (2003). Socioeconomic Status Modifies Heritability of IQ in Young Children. *Behavior Genetics*

Twain, M. (1869). *Innocents Abroad or the New Pilgrim's Progress.* American Publishing Company. NY, N.Y.

U.S. Department of Agriculture, Food and Nutrition Service. (2010.) Income eligibility guidelines. Retrieved February 24, 2011 from http://www.fns.usda.gov/cnd/Governance/notices/iegs/IEGs.htm

U.S. Department of Health and Human Services, Assistant Secretary for Planning and Evaluation. (2011.) Poverty guidelines, research, and measurement. Retrieved February 24, 2011 from http://aspe.hhs.gov/poverty/index.shtml

Victora, C. G., Bahl, R., Barros, A. J. D., França, G. V. A., Horton, S., Krasevec, J., Murch, S.,

Sankar, M. J., Walker, N., Rollins, N. C., for The Lancet Breastfeeding Series Group. (2016). Breastfeeding in the 21st century: epidemiology, mechanisms, and lifelong effect. *Lancet.* Volume 387, Issue 10017, 30 January–5 February 2016, Pages 475-490

Willerman, L. Naylor, A. F., and Myrianthopoulos, N. C. (1970). Intellectual development of children from interracial matings. *Science*, 170, 1329-1311

Willerman, L., Naylor, A. F., and Myrianthopoulos, N. C. (1974). Intellectual development of children from interracial matings: Performance in infancy and at 4 years. *Behavior Genetics*, 4, 84-88.

Wines, Michael. December 28, 2006. Malnutrition Is Cheating Its Survivors, and Africa's Future. NY Times

Zajonc, R., and Bargh, J. (1980). "The Confluence Model: Parameter Estimation of Six Diveredent Data Sets on Family Factors and Intelligence." *Intelligence*, 4, 349-361.

About the Author

While working at IBM for 34 years focusing on the development of advanced system designs, Thomas Schick published many patent disclosures and journal articles. During his years with IBM he also earned an MS degree from Duke University. In addition, while at IBM, he mentor-tutored about 50 learning-disabled children, always one-on-one. After retiring, and while continuing his activities with these children, he pursued a PhD in cognitive psychology. With those experiences and an additional 15 years of in-depth research, he wrote this book. Tom and Anne, his wife, live in Cary, North Carolina. They have two children and four grandchildren.

www.ingramcontent.com/pod-product-compliance
Lightning Source LLC
Chambersburg PA
CBHW030002190526
45157CB00014B/112